T0342250

Explosion Vented Equipment System Protection Guide

Robert C. Comer, P.E.

This edition first published 2021

Registered Office
John Wiley & Sons, Inc., 111 River Street, Hoboken, NJ 07030, USA

Editorial Office
111 River Street, Hoboken, NJ 07030, USA

For details of our global editorial offices, customer services, and more information about Wiley products visit us at www.wiley.com.

Wiley also publishes its books in a variety of electronic formats and by print-on-demand. Some content that appears in standard print versions of this book may not be available in other formats.

Library of Congress Cataloging-in-Publication Data

Names: Comer, Robert Charles, 1934– author.
Title: Explosion vented equipment system protection guide / Robert Charles Comer.
Description: First edition. | Hoboken, NJ, USA : John Wiley & Sons, 2020. | Includes index.
Identifiers: LCCN 2020022867 (print) | LCCN 2020022868 (ebook) | ISBN 9781119640035 | ISBN 9781119640066 (adobe pdf) | ISBN 9781119640073 (epub)
Subjects: LCSH: Dust control–Equipment and supplies. | Fire prevention–Equipment and supplies.
Classification: LCC TD884.5 .C655 2020 (print) | LCC TD884.5 (ebook) | DDC 628.9/22–dc23
LC record available at https://lccn.loc.gov/2020022867
LC ebook record available at https://lccn.loc.gov/2020022868

Cover design : Wiley
Cover Images: (background) © Jose A. Bernat Bacete/Getty Images
(graph) Courtesy of Robert Comer

Set in 9.5/12.5pt STIXTwoText by SPi Global, Pondicherry, India

This book is dedicated to my beautiful, extraordinary wife Jean and the beautiful, talented family that we have been blessed with. Daughter Lynn, her husband Jack, their son Adam, his wife Jessica, their son Andrew, his wife Sarah; daughter Donna, her husband George, their son Jonathan, their daughter Jacqueline; and daughter Kerry Ann, her husband John, their daughter Samantha, their sons Christopher and Stephen.

Contents

Preface

This book presents practical applications and guidance for engineers. It provides a reference to meet the needs of a company licensed or competent unlicensed engineer, that by education or experience understands the concepts presented in this book. It provides guidance to analyze and design or retrofit dust collection equipment to resist dust explosions, to protect employees and reduce production down time. The intent of the book is not to encourage licensed or competent unlicensed engineers to practice outside of their field of expertise, but to provide guidance to competent engineers to enhance their knowledge and approach to the narrow area of dust handling explosion venting and system reinforcing. Dust collector systems are not very complicated in scope and design. The reinforcing of thin panels is common to all square/rectangular dust collectors as are cylindrical dust collector roofs and access doors. Bolted flange connections, hinges, and latches are common and easily analyzed. This book provides guidance for these conditions. Licensed engineers and engineering consulting firms may find this book to be an advantage in obtaining the business of mitigating hazards for the many facilities that will have their hazards exposed by the National Fire Protection Association (NFPA) mandate of providing a dust hazard analysis of all facilities by September 2020. Many facilities do not have an engineer on staff, and they will be obligated to hire an outside consultant to mitigate their hazards.

Detailed explanations of formulas and tedious derivations are not necessary and are avoided. The data are presented in tables and graphs along with examples to illustrate the actual applications.

For each topic, the general principles and theories are stated, followed by extensive tables and worksheets for use in calculations of stress and strain and to design reinforcing of equipment with adequate and economical reinforcing members.

The examples are based on actual proven designs developed by the author by analyzing over 200 systems to clearly illustrate application of the information provided.

The book is arranged to provide a means of solving practical engineering problems.

Although every effort has been made to avoid errors, it is possible some could exist. I will be grateful for any needed corrections.

Robert C. Comer
Rockaway, NJ
09 June 2020

Introduction

Be aware: Your dust collection system, though properly vented per National Fire Protection Agency, NFPA 68 (2013) "Standards on Explosion Protection by Deflagration Venting," may not survive the vented explosion without having permanent deformation or catastrophic failure allowing hot gasses or shrapnel to be expelled into the surrounding area causing personnel injury or death, and exceedingly long production lost time while new equipment is ordered or failed equipment is rebuilt. **The new NFPA 652 "Standard on the Fundamentals of Combustible Dust," applies to all agriculture and food production facilities**. All facilities must complete a Dust Hazard Analysis (DHA) by September 2020. Required is a systematic review to identify and evaluate potential dust fire, flash fire, and explosion hazards in a process or facility where combustible/explosible material is handled or processed. There are over 130 000 plants that handle sugar, flour, starch, dried milk, egg whites, gluten, and artificial sweeteners and are dealing with combustible dusts and potential dust explosions. Milling, grinding, spray drying of liquids, and handling of grains are processes that generate combustible dust. Combustible dusts are common in the food, plastics, metals, agriculture, chemical, and wood industries. There are many documented cases of dust explosions causing injury and death to workers due to a lack of understanding of the equipment requirements. Many plants have been in operation for years without an explosion and without being cited by OSHA. This does not mean that they are in compliance, and their facility processes will not injure or cause death to employees.

The expense of retrofitting equipment is justified when, in the event of a dust explosion, there is major production down time, or employees have not been protected properly. A recent minor dust explosion in a 3D printing company caused third degree burns on an employee. The fine by OSHA was US$ 64 500. Proper equipment reinforcing design would have cost a small fraction of that fine. The US Chemical Safety Board reports 316 dust explosions over the last 30 years that caused 145 workers killed and 846 injured with extensive damage to facilities.

Damage to facilities and lost production time has been extensive. The most deadly 2017 event occurred when corn dust exploded at a milling plant in Wisconsin, killing 5 and injuring 14. OSHA has levied a fine of US$ 1.8 million citing 19 violations at the mill. According to the independent 2019 (mid-year) Combustible Dust Incident report from DustEx Research Ltd., there were 80 dust-related fires, 19 dust related explosions, 22 injuries, and 1 fatality in the United States from January through July 2019. This could have been avoided with proper design of the equipment.

This manual provides the design criteria and guidance to ensure safe venting of dust collection systems. Purchasers of dust collection systems as well as manufacturers of the equipment can ensure that the equipment is not only vented safely and correctly but also would not contribute to lost production time. The equipment would be able to be put back in service with only a cleaning and replacement of the dust explosion relief elements. Analysis performed by a licensed engineer consultant would cost US$ 8000.00 or more for each system. This book will allow a licensed or competent unlicensed engineer, that by education or experience, understands the concepts presented in this book. It provides guidance to analyze, design, and supervise the installation or retrofit reinforcing of dust handling equipment. This manual also provides analysis and design of the explosion relief ductwork required to vent the hot gasses outside to a safe location.

Note: An alternative to the explosion relief ductwork is the use of a device called a flameless vent that is installed over the standard explosion vent to extinguish the flame front as it exits the vented area. This allows the venting to be inside without a duct to the outside when access to the outside is remote or venting to a safe location is not possible. Flameless devices are not recommended for toxic dusts because dust can be released into the room. Consult the manufacturer for safe application of this device.

It is essential that no catastrophic failure of the equipment occurs that could injure an employee working in the area by hot gasses or shrapnel escaping from the equipment. If hot gasses or shrapnel escapes from the equipment any dust accumulated on the floor, beams, or other equipment in the area due to poor housekeeping would be blown into the air where it forms an explosive mixture that becomes a secondary explosion, usually of higher intensity than the first explosion. Extensive building and equipment damage and personnel injury or death would occur.

Dust collection equipment is normally designed and certified for vacuum (negative pressure) service by the manufacturer; however, when the equipment is subjected to an internal positive pressure due to the vented explosion, P_{Red}, there is the possibility of structural failure. A properly vented system as per NFPA 68 (2013), "Standards on Explosion Protection by Deflagration Venting," using the deflagration parameters of the combustible dust-air mixture found by testing in accordance with ASTM E1226-10, "Standard Test Method for Explosibility of Dust Clouds," results in a vented explosion flowing pressure, (P_{Red}), in the equipment that must be contained safely. It has been my experience in analyzing over 200 installations worldwide that the equipment must be reinforced in almost every case to safely contain the vented explosion flowing pressure (P_{Red}). The explosion relief vent element release pressure (1.0 ± 0.25 psig minimum) is usually higher than the equipment allowable pressure ($\pm 20''$ water gauge, 0.723 psig). The cost of reinforcing is usually less expensive than the cost of equipment panels being made of thicker material to contain P_{Red}, or to replace the equipment.

Part 1: Structural Analysis and Design for Reinforcing Dust Handling Systems

The analysis in this manual is presented to allow a licensed or competent unlicensed engineer, that by education or experience understands the concepts presented in this book, to determine the required reinforcing design when reviewing the equipment supplied by the manufacturer.

All he/she needs to know is the explosion vent size, the vented explosion flowing pressure, P_{Red}, and the equipment wall material and thickness.

An equipment manufacturer can also use this manual to design their equipment to sustain any pressure dictated by the user after a venting analysis is performed, an explosion relief element is decided on and P_{Red} is determined.

Dust collectors are either cylindrical or square/rectangular shaped. Cylindrical dust collectors are usually structurally sound except for those with a flat roof and an access door that may require reinforcing. Most dust collectors and associated equipment have large, flat, thin panel construction that is very easy to analyze for reinforcing. A weak link in both types of construction is the roof, access door, and door restraints (hinges and latches). In almost every case, the access door and its restraints must be reinforced so that it does not fail and allow hot gas or shrapnel to be discharged into the surrounding area where employees may be working, or other equipment may be damaged. A secondary greater explosion of dust on surfaces in the area by poor housekeeping could occur that could destroy the building structure. Flow activated explosion isolation valves and rotary valves installed in the equipment upstream and downstream of the dust collector contain the combustible dust deflagration products, smoke, dust, burning debris from traveling beyond the equipment.

Part 2: Explosion Relief Element and Explosion Flowing Pressure Analyses

This section of the book is presented to allow a competent engineer that by education or experience understands the concepts presented in this book, to determine the explosion relief element to safely vent and lower the explosion flowing pressure, P_{Red}, to a reasonable level so that reinforcing, if required, can be analyzed and designed in a practical, economical manner as defined in Part 1.

Required for this section is the process dust explosion characteristics, K_{st}, P_{max}, and the dust handling system geometry. If the dust explosion characteristics are not known, a sample of the dust must be tested to obtain these characteristics. The table of dust characteristic values presented in this book is for estimating purposes only. To ensure proper design, a tested sample of dust must be used.

How to Use This Book

There are two parts to this book. Part I provides the structural analysis and design for reinforcing of dust handling equipment and explosion relief ducting to prevent a catastrophic failure of the equipment during a dust explosion. Part II provides the analysis of dust handling equipment to determine the size of the explosion relief burst element and the explosion relief flowing pressure (P_{Red}) based on the process dust explosion characteristics and geometry of the equipment.

Existing in-house dust collection systems or new proposed equipment: If the existing or new equipment has been certified by the manufacturer for the allowable safe operating positive pressure on the drawings or in the specifications, it must be determined from the manufacturer, the design criteria that was used (% of yield or ultimate stress). If 2/3 of the 0.2% yield strength was not used as the limiting stress, the equipment does not comply with NFPA 68 (2013) "Standards on Explosion Protection by Deflagration Venting." Part 1 of this book will provide the guidance and structural analysis to ensure that the equipment complies.

If the equipment has been designed to be explosion vented and has an explosion vent built-in or proposed, there must have been an explosion venting analysis performed specific for the process dust being handled and geometry of the system. If an analysis is provided with the resultant explosion flowing pressure (P_{Red}), then go to Part 1 directly to ensure the integrity of the system. If the manufacturer does not provide the analysis, Part 2 of this book will provide the guidance and analysis to size the explosion relief burst element, ducting and to calculate the explosion flowing pressure (P_{Red}) before accessing Part 1 to ensure that the system is structurally sound and safely vented.

Symbols

A_d	shear area at minor diameter of bolt, in.^{2}
A_{door}	cross-sectional area of access door, in.^{2}
A_p	cross-sectional area of panel, in.^{2} = $W_p t_p$
A_{pin}	area of pin, in.^{2}
A_s	shear area, in.^{2}
A_v	cross-sectional area of vent and minimum annular perimeter area of cage, in.^{2}
A_w	weld area, in.^{2}
$A_{1,2,3,4}$	nozzle and reinforcing ring areas, in.^{2}
a	length of the longest side of the panel, in.
b	length of the shortest side of the panel, in. (*Note*: The short side of the panel is always the controlling side in the stress analysis.)
bar	14.50 psi
C	distance from centroidal axis to panel center, in.
C_p	$t_p/2$
DLF	dynamic load factor = 1.2
E	Young's modulus of elasticity in tension and compression, psi (for steel this value is 30×10^6 psi)
e	efficiency, %
F	axial tension in cylinder wall, lbs/in.
F_a	allowable load per pair of welds, lbs
f_s	shear load on bolts, psi
f_{1max}	maximum stress at center of panel long edge, psi
f_{2max}	maximum stress at center of panel, psi
f'	combined shear and tension in bolts, psi
f_{max}	maximum stress, psi
F_T	total load on bolts, lbs
F_{tu}	ultimate strength of material, psi
F_{ty}	0.2% yield strength of material, psi

F_r	vented reaction force of explosion, lbs
F_x	horizontal reaction force of vented explosion, lbs
F_y	vertical reaction force of vented explosion, lbs
G	modulus of elasticity in shear, psi
h	cylinder wall thickness, in.
H	height of member, in.
I_C	moment of inertia of composite member, in.^{4}
I_P	moment of inertia of panel, in.^{4}
K_{st}	explosibility of material, bar-m/s
L	length, span of beam, etc., in.
L_B	load per bolt, lbs
L_T	total load on flange, lbs
L_t	total load on each panel, lbs
L_c	concentrated load, lbs
L_w	weld length, in.
M_{max}	maximum moment, in.-lbs
M_o	edge moment on panel, in.-lbs
N	number of bolts
P_{max}	maximum pressure of deflagration, bar
P_{red}	explosion flowing pressure, psi
P_{stat}	vent burst element release pressure, psi
Q	statical moment of panel about centroidal axis = A_pC, lbs/in.
R_{avg}	average reaction force normal to panel surface, lbs/in. ($a \times b \times P_{red}/2a + 2b$)
R_{max}	reaction force normal to the panel surface, lbs/in. (at center of long side)
$R_l = R_r$	end reactions on panels and reinforcing members, lbs/in.
R_o	cylinder radius, in.
S_a	allowable stress (2/3 of the 0.2% yield strength), psi
S_c	Section modulus of reinforcing composite member, in.^{3}
S	stresses in material, psi
S_u	ultimate strength of the material, psi
t_p	thickness of the panel, in.
t_w	weld size, in.
T	bolt torque, in.-lbs
T_t	total load on panel, lbs
T_p	thickness of reinforcing ring, in.
v	Poisson's ratio
V_o	shear load on panel edge, lbs
V_m	maximum vertical shear end reaction load, total load/2, lbs

V_h	horizontal shear between panel and reinforcing member, $= V_m Q/I_c =$ lbs/in.
w	uniform edge load acting on the reinforcing member, lbs/in. $= R_{max} \times 2$ panels
w.g.	water gauge pressure, inches of water, psi
W_b	maximum individual bolt load, lbs
W_p	width of panel or reinforcing ring, in.
wt/ft	weight per foot of reinforcing member, lbs/ft
y_{max}	maximum deflection of the panel, in.
Z	distance of bolt farthest from neutral axis, in.

Greek Symbols

α	coefficient from Table 2.1
β	coefficient from Table 2.1
Δ	growth of cylinder under load, in.
$\sigma_{x,y}$	stress in cylinder wall, psi
σ_{max}	stress in head, psi

Note: All pressures are gage pressures unless otherwise indicated.

Typical Dust Collection System Checklist

Example (Rectangular Dust Collector)

System Identification: Typical dust collector example
Process Product: No specific product assumed
Product P_{max}__________ Product K_{st}__________ P_{red} = 5.0 psi assumed
Basis of P_{max} and K_{st}: Was product dust tested ____Y ____X ____N
Were values approximated from a chart ____Y ____X ____N
Other: Average values assumed for this example

Dust collector: If square or rectangular

Pressure rating from manufacturer: ±20 in water gauge
Basis for rating: 60% of 0.2% yield strength, or other__________
Is vessel located inside or outside? Inside
Top width: 97.75 in. Depth: 38.25 in. Height: 27.75 in.
Top sheet thickness: 10 ga. (0.1345 in.)
Sides height: 108.25 in.
Sidewall thickness: 10 ga. (0.1345 in.)
Reinforcing on top or walls: None
Volume of dust collector: N/A

Dust collector: If cylindrical

Pressure rating from manufacturer:__________
Basis for rating: ________% of 0.2% yield strength, or other________
Diameter: ________
Type of top: Flat______ Ellipsoidal__________ Torispherical________
Thickness of top: ________
Thickness of cylinder wall: ________
Length of cylinder: ______________
Volume of dust collector: ______________________________

Hopper height: 30.50 in. Discharge size: 9.25 in. × 9.25 in.
Hopper wall thickness: 10 ga. (0.1345 in.)
Dust collector/hopper flange size and thickness: 2″ × 2″ × 1/4″ angle
Bolt size and spacing: ______________________
Volume of hopper: N/A

Number of bags/cartridges: N/A
Bag/cartridge diameter: __________ Number of bags/cartridges: ______
Bag/cartridge volume: ______________

Access door width and height: 37.875 in. × 40 in.
Access door thickness: 10 ga. (0.1345 in.)
Reinforcing on door: None
Type of latching:
Clamps________________ Bolts ________________
Hinges ____________ Hinge pin diameter ______________

Nozzles: Diameter ______________________
Dusty air inlet dimensions: ______________________
Clean air outlet dimensions: ______________________
Leg supports:
Size: ______ Number of supports: 4 Length of supports: 78.50 in.
Footpad size and thickness: 7.50 in. × 7.50 in. × 1/2 in.
Attachment to floor or ground: 3/4 in bolts to floor

Explosion vent information:

Manufacturer/part number: N/A
Size: Round _______ Square/rectangular: 24 in. × 24 in.
Vent pressure rating: 1.0 psi ± 0.25 psi Vent area: ______________
Flange size: 2″ × 2″ × 1/4″ angle Bolt size and number
Distance from top of dust collector: 35 in.

Duct information:

Size: 24 in. × 24 in. Length: Wall thickness: 10 ga. (0.1345 in)
Any bends at end of duct?

Screw conveyor:

Cover width __________ Cover length _________ Thickness _________
Spacing and size of bolts or clamps on cover:

Acknowledgments

It is a pleasure to acknowledge my appreciation to people that inspired this manuscript. I am particularly indebted to George Petino, Jr., MME, P.E., Dust Explosion Consultant and owner of Hazards Research Corporation, who provided me with the opportunity to consult with facilities worldwide in mitigating their dust explosion hazards. His professional mentoring and guidance has been extraordinary.

I am also indebted to my son-in-law Jack Kehlenbeck who aided in compiling the manuscript, his computer skills are outstanding.

NFPA, National Fire Protection Association, must be applauded for their work in making and keeping the Standards on Explosion Protection by Deflagration Venting viable and progressive.

I must thank John Wiley & Sons, Inc., Project Editors Ms Beryl Mesiadhas, Michael Leventhal, Summers Scholl, and Vishnu Priya for their excellent professional guidance.

Finally, as always, I could accomplish nothing without the support and encouragement of my wife Jean.

Part 1

Structural Analysis and Design for Reinforcing Dust Handling Systems

Part 1: Introduction

This part provides the structural analyses for the reinforcing of the dust handling system components and the design of the ductwork and weather covers associated with the explosion venting element to provide a safe discharge of the hot gasses to a safe location. This part applies to existing equipment already installed in a facility or to new equipment on order or to be ordered. The stability of the dust collectors is addressed when subjected to the explosion venting forces. The environment surrounding the explosion venting is addressed to preclude exposure to personnel or equipment. Electrical grounding design of components is detailed to preclude static discharge that may cause ignition of dust. General housekeeping, a very important aspect of keeping the area safe from a potential explosion, is also addressed. The Appendix A to Part 1 includes worksheets to follow in the structural analyses.

This part can only be used before Part 2 if the following design criteria are available:

1) Drawings of the system with materials of fabrication, dimensions, and wall thicknesses.
2) The process material is stated with test results of the explosibility of the material, (K_{st}), and the reduced explosion flowing pressure (P_{Red}), that the system will be exposed to in the event of an explosion. It is important to note that new tests may be required of the material being processed if the process material is changed at any time. If the process material is changed to a more highly explosive material, it may be necessary to reanalyze the equipment to ensure that the explosion vent and reinforcing is adequate.
3) An explosion vent has been sized and its location on the system is established.
4) The location of the system equipment is established, either in the facility or outside of the facility, and the surrounding area is defined to ensure that the explosion vent will not endanger any personnel, equipment, or building structure.

If all of the listed criteria required to use Part 1 is not available, you must go to Part 2 to use the system parameters to analyze and design the explosion vent size to ensure that a safe reduced explosion flowing pressure, (P_{Red}), is provided.

Explosion Vented Equipment System Protection Guide, First Edition. Robert C. Comer.

1

Design Criteria and General Theory

Figure 1.1 illustrates the pressure versus time of an explosion event. An unvented explosion reaches a very high pressure and would result in a catastrophic failure of the system components. The explosion must be vented to reduce the pressure. Ideally, the vented explosion flowing pressure would be below the strength of the equipment; however, most dust collectors and explosion vented systems are quite weak and require reinforcing to resist a flowing pressure obtained with a practical vent and duct size. As shown on Figure 1.1, the explosion vent does not open until the explosion venting pressure exceeds the allowable pressure of the equipment.

After the explosion venting analysis has been performed on the equipment in accordance with the guidelines set forth in NFPA-68 (2013). The explosion relief element has been sized and a resultant P_{Red} (explosion flowing pressure) has been determined. P_{Red} is the design pressure to be used in this structural analysis. Permanent deformation of the equipment when subjected to an explosion is not acceptable in a production environment, where it is desirable to put the equipment back in service as soon as possible. NFPA-68 recommends that 2/3 of the 0.2% yield strength of the material is used as the design criteria. This is the limiting stress that will ensure that no permanent deformation will occur. With the conservative approach used in this stress analysis there are built-in extra factors of safety. The reinforcing and welding costs are greatly reduced to a practical level without sacrificing safety. This will also ensure that no catastrophic failure can occur. Catastrophic failure will occur if the stress exceeds the ultimate strength of the material.

The equipment may already be installed and in use in your facility or new equipment is to be, or is, on order. It is important that the manufacturer's drawings and specifications of the equipment are reviewed for allowable normal operating design pressure and an analysis performed to ensure safety. The panel thicknesses

Explosion Vented Equipment System Protection Guide, First Edition. Robert C. Comer.

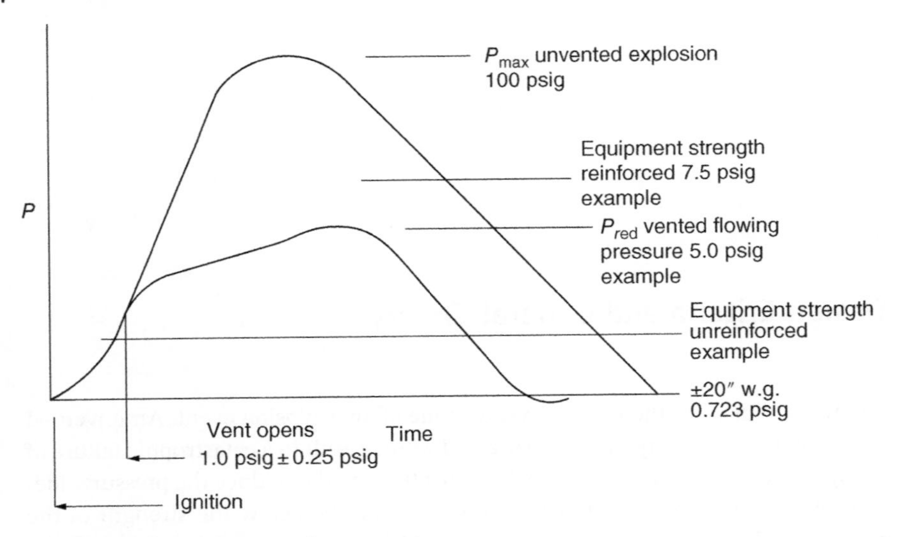

Figure 1.1 Pressure versus time example.

and the working design pressure should be noted on the drawings or in the specifications. If all of the information is available except for the material thicknesses, there are wall thickness detectors available that can be used to determine the existing wall thicknesses where access is limited to one side. These ultrasonic thickness detectors can be used to evaluate the amount of corrosion or erosion that exists also. A reliable detector that is available that will read the wall thickness even through a paint coating is DeFalsko Co., Def-Positector Model UTG-ME, mode multiple E. This unit has a range of 0.040 in. to approximately 5.0 in. wall thickness with an accuracy of 0.0011 in. for carbon and stainless steel. Another reliable ultrasonic detector is Olympus Co., Model 26MG, also usable where limited to one side. This unit has a range of 0.020–19.999 in. wall thickness with an accuracy of 0.001 in. Chapter 2 provides an evaluation of the noted vessel working pressure by the manufacturer, and why it is usually inadequate to prevent a catastrophic failure. The equipment is fabricated from carbon steel, stainless steel, or galvanized steel. The reinforcing members are to be plates, structural angles, or structural tubing. Refer to Table 1.1 for allowable stresses. Refer to Table 1.2 for material versus gauge.

As described in the Part 2 explosion venting analyses, the pressure change during a dust explosion is too slow to consider a dynamic stress analysis where the material allowable stress properties are higher. A static pressure analysis is used throughout the book and this also provides a greater safety factor.

Table 1.1 Material properties and allowable stresses.

Material properties		
Structural steel shapes ASTM A36	$F_{ty} = 36\,000$ psi	$F_{tu} = 58\,000$ psi
Structural steel tubing A500, Grade B	$F_{ty} = 46\,000$ psi	$F_{tu} = 58\,000$ psi
Steel sheet SAE 1020	$F_{ty} = 32\,000$ psi	$F_{tu} = 55\,000$ psi
Galvanized steel sheet ASTM A526/527	$F_{ty} = 33\,000$ psi	$F_{tu} = 56\,000$ psi
Stainless steel sheet (304)	$F_{ty} = 35\,000$ psi	$F_{tu} = 85\,000$ psi

Allowable stresses: Two-third (0.67) of the 0.2% yield strength is the limiting stress if no deformation is allowed. If deformation is allowed, then 2/3 of the ultimate strength of the material is the limiting stress per NFPA-68. To reduce production down-time no deformation is allowed.

With no deformation allowed	
Structural steel shapes	Sa = 0.67 (36 000) = 24 120 psi
Structural steel tubing	Sa = 0.67 (46 000) = 30 820 psi
Steel sheet	Sa = 0.67 (32 000) = 21 440 psi
Galvanized steel sheet	Sa = 0.67 (33 000) = 22 110 psi
Stainless steel sheet	Sa = 0.67 (35 000) = 23 450 psi

Source: Data from ASTM Metals Handbook.

Table 1.2 Steel sheet gauges.

Gauge	Carbon steel	Stainless steel	Galvanized steel
12	0.1046	0.1054	0.1084
11	0.1196	0.1200	0.1233
10	0.1345	0.1350	0.1382
8	0.1644	0.1650	0.1681
7	0.1793	0.1874	—
3/16	0.1875	0.1875	—
1/4	0.2500	0.2500	—
5/16	0.3125	0.3125	—
3/8	0.3750	0.3750	—
1/2	0.5000	0.5000	—

It must be kept in mind that the material dust being processed now may not be the material dust processed in the future. A change in material processed must be evaluated for explosibility and the explosion vent changed if the dust is more dangerous.

2

Square/Rectangular Dust Collector Wall, Roof, and Hopper Sections

Figure 2.1 illustrates a typical square/rectangular dust collector configuration. As noted on the drawing, the manufacturer has specified that the housing is rated for ±20″ w.g., a normal pressure rating for a dust collector in vacuum service. The equivalent pressure is ±0.723 psig. Explosion venting elements, burst disks, etc., are rated to burst at a minimum of 0.5 ± 0.3 psig and can range up to 10 ± 1.0 psig. As the burst element releases the hot gasses the pressure increases rapidly to the vented explosion flowing pressure, P_{Red} value. This value will always exceed the 0.723 psig rated value of the dust collector; thereby, requiring reinforcing of the dust collector. For the examples, the P_{Red} value used in this manual will be 0.36 barg (5.0 psig), a value that is the average P_{Red} for over 200 analyses in my experience. The range of P_{Red} has varied from 0.15 barg (2.1 psig) to 0.6 barg (8.4 psig). The wall thickness for the dust collector used in the examples is 10 ga. (0.1345″) for carbon steel sheet, the most common material used, and also the average wall thickness for most dust collectors. Refer to Table 1.2 for steel material and sheet thicknesses versus gauge. Substitute the material and wall thickness of your equipment and substitute the vented explosion flowing pressure, P_{Red}, calculated by others in accordance with NFPA-68 for your system. *Warning*: Any dust collection vessel with a design pressure, $P_{Red} = 15$ psig or higher (very rare) is considered a high pressure vessel, and it must be designed and built according to the codes and standards of the American Society of Mechanical Engineers (ASME) International. You must work with your dust collector supplier to design and build a high-pressure vessel that complies with the ASME Code.

Refer to Table 1.1 for material versus allowable stresses.

Explosion Vented Equipment System Protection Guide, First Edition. Robert C. Comer.

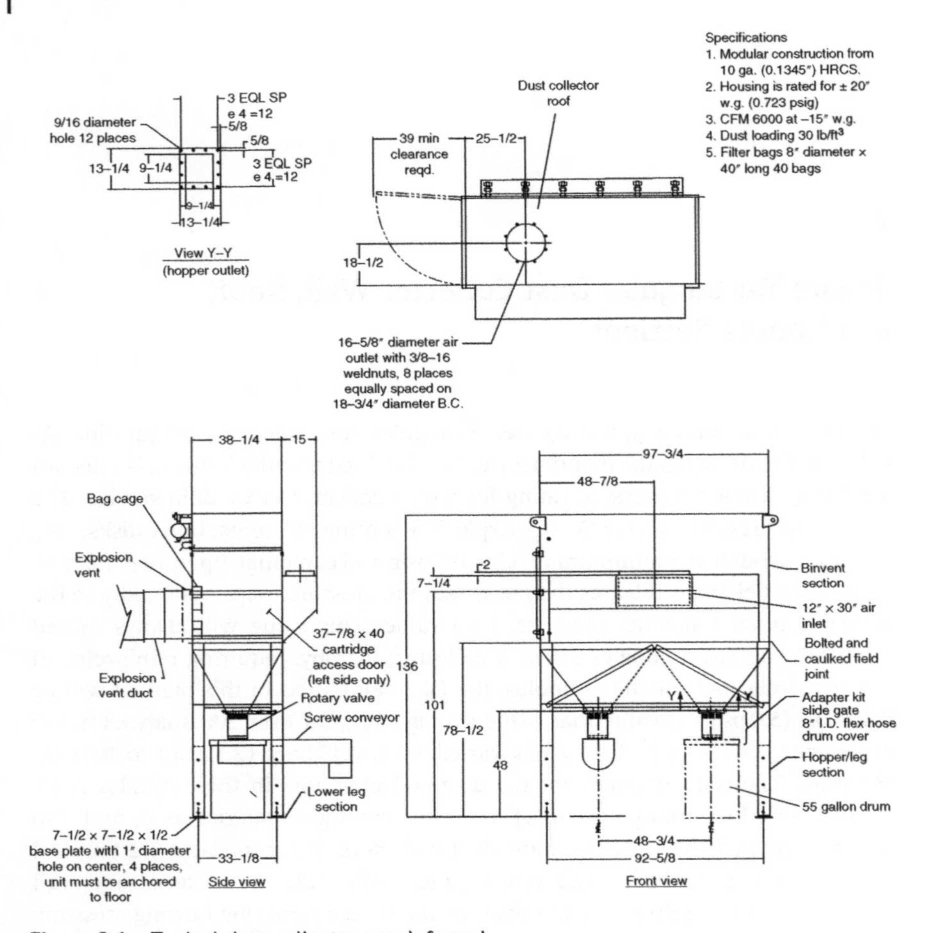

Figure 2.1 Typical dust collector unreinforced.

Figure 2.2 illustrates the reinforcing of the sample dust collector. The details of the reinforcing are shown on figures in the text.

Rectangular Panel Stresses

The panel stresses are obtained using the formulas for "f_{max}," where the panel has all edges fixed.

The panel deflection under load is determined by the deflection formula for "y_{max}."

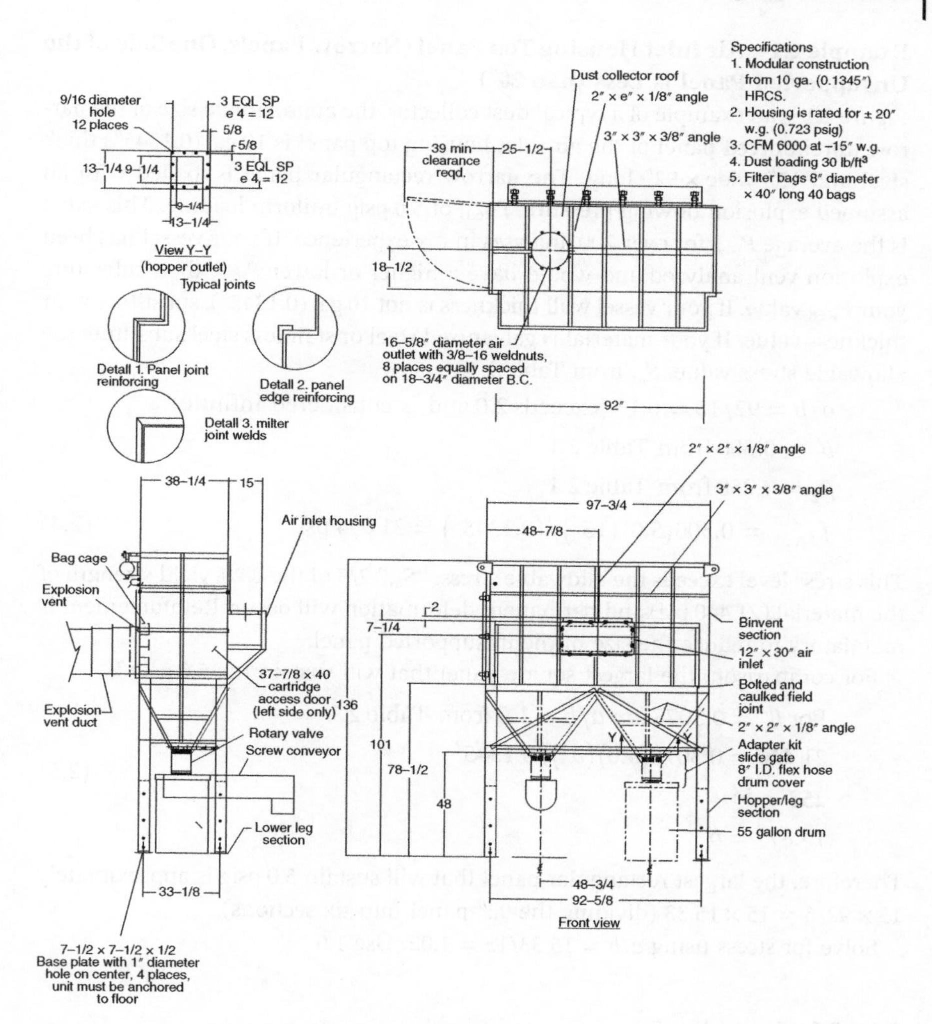

Figure 2.2 Typical dust collector reinforced.

If the deflection of the panel under load exceeds half of the panel thickness, $t/2$, then panel diaphragm stresses may be considered. A discussion of diaphragm stresses is given later in the chapter.

$$f_{1\,\max} = \beta_1 P_{\text{Red}} b^2 / t_{\text{p}}^2 \text{ at center of long edge} \tag{2.1}$$

$$f_{2\,\max} = \beta_2 P_{\text{Red}} b^2 / t_{\text{p}}^2 \text{ at center of panel} \tag{2.2}$$

$$y_{\max} = \alpha P_{\text{Red}} b^4 / E t_{\text{p}}^3 \tag{2.3}$$

Example 2.1 Air Inlet Housing Top Panel (Narrow Panels, One Side of the Unsupported Panel is Less than 20″)

Figure 2.1 is an example of a typical dust collector, the approximate size of the narrow unreinforced panel of the air inlet housing top panel is 10 ga. (0.1345″) thick steel and 15″ wide × 92″ long. The narrow rectangular panel is loaded with an assumed explosion flowing pressure, P_{Red}, of 5.0 psig uniform loading. This value is the average P_{Red} for over 200 analyses in my experience. If your vessel has been explosion vent analyzed and would have a higher or lower P_{Red}, then substitute your P_{Red} value. If your vessel wall thickness is not 10 ga. (0.1345″), substitute your thickness value. If your material is galvanized steel or stainless steel substitute the allowable stress value, S_a, from Table 1.1.

$$a/b = 92/15 = 6.13 \text{ (exceeds 2.0 and is considered infinite)}$$

$$\beta_1 = 0.500 \text{ from Table 2.1}$$

$$\beta_2 = 0.250 \text{ from Table 2.1}$$

$$f_{1\,max} = 0.500(5.0)(15^2)/(0.1345^2) = 31\,094 \text{ psi} \tag{2.4}$$

This stress level exceeds the allowable stress, "S_a," 2/3 of the 0.2% yield strength of the material (21 440 psi) and permanent deformation will occur. Reinforcement is mandatory to reduce the size of the unsupported panel.

For comparison, the largest square panel that will sustain the 5.0 psig is

$$\begin{aligned} &\text{For } \beta_1 = 0.3078, \text{ for } a/b = 1.0 \text{ from Table 2.1} \\ &21\,440 = 0.3078(5.0)(b^2)/0.1345^2 \\ &252 = b^2 \\ &15.87'' = b \end{aligned} \tag{2.5}$$

Therefore, the largest rectangular panel that will sustain 5.0 psig is approximately 15 × 92/6 = 15 × 15.33 (dividing the 92″ panel into six sections).

Solve for stress using a/b = 15.33/15 = 1.02. Use 1.0.

Table 2.1 Panel edges fixed.

a/b	1.0	1.2	1.4	1.6	1.8	2.0	>2
β_1	0.3078	0.3834	0.4356	0.4680	0.4872	0.4974	0.500
β_2	0.1386	0.1794	0.2094	0.2286	0.2406	0.2472	0.250
α	0.0138	0.0188	0.0226	0.0251	0.0267	o.0277	0.0284

Values of a/b that exceed 2.0 are considered infinite and $\beta_1 = 0.500$, $\beta_2 = 0.250$, and $\alpha = 0.0284$.

Source: Roark's Formulas for Stress and Strain, Warren C. Young, McGraw-Hill Book Co., 6th Edition, Table 26, Case 8a, page 465. Reproduced with permission of McGraw-Hill Book Co.

From Table 2.1, for $a/b = 1.0,\ \beta = 0.3078$

$$f_{1\,\max} = 0.3078(5.0)(15^2)/0.1345^2 = 19\,141 \text{ psi} \tag{2.6}$$

This stress is within the allowable stress of 21 440 psi. The actual panel size is approximately 3″ less in width and length when a reinforcing angle or tube is welded to it (see Figure 2.5).

The actual panel stress in between reinforcing members is then $a/b = 12.33/12 = 1.022$.

Still using 1.0

From Table 2.1, for $a/b = 1.0$

$$f_{1\,\max} = 0.3078(5.0)(12^2)/0.1345^2 = 12\,251 \text{ psi} \quad \text{(well within the allowable stress of 21 440 psi)} \tag{2.7}$$

The 15″ × 92″ long panel must be divided into six sections with reinforcing members.

Check the deflection of the panel:

From Table 2.1, for $a/b = 1.0$

$$y_{\max} = 0.138(5.0)(12^4)/30 \times 10^6(0.1345^2) = 0.0196 \text{ in.} \tag{2.8}$$

The deflection of the panel does not exceed half the panel thickness.

When the deflection of the panel exceeds half the panel thickness, the middle surface of the panel becomes appreciably strained, and the stress is called diaphragm stress. This condition allows the panel to carry part of the load in direct tension balanced by the radial tension at the edges of the panel. The panel is then stiffer and a given load produces less stress than ordinary theory.

By ignoring a complicated diaphragm stress analysis and analyzing the panels with fixed edges, the resulting calculated stress is higher than actual and provides a greater safety margin.

If an advanced diaphragm stress analysis of a panel with a deflection greater than half the thickness of the panel is desired, refer to Young (1989).

All large panels on the unit must be checked for stress and reinforced. Usually, the spacing resulting from an analysis such as Examples 2.1 and 2.2 can be used on all sides. The reinforcing members must be sized accordingly.

Reinforcing Members Analysis for Example 2.1

Reinforcing members that are best suited are ribs (Table 2.2), angles, channels, and structural tubing.

Angles placed with the both legs against the panel provide a cleaner surface that sheds dust and is the preferred member; however, square or rectangular structural tubing or channels are to be used if a flat surface is needed; for example, access door reinforcing where latches and hinges are required to be attached to a flat surface (refer to access door latch and hinge Chapter 5). Table 2.3 presents the structural characteristic, S_c, section modulus, of various angles combined with a portion of the panel as a composite section. It is recommended that 2″ × 2″ × 1/8″ angles are the smallest members to be considered. This size is readily available in stock. It is also recommended that an angle larger than 5″ × 5″ is not used as the panel under the angle becomes too wide and unsupported.

Table 2.4 presents the structural characteristic, S_c, section modulus, of various square and rectangular tubing combined with a portion of the panel as a composite section. It is recommended that 2″ × 2″ × 3/16″ thick tubing is the smallest member to be considered. This size is readily available in stock.

Table 2.5 presents the structural characteristic, S_c, section modulus, of various channels combined with a portion of the panel as a composite section. It is recommended that C3–4.1 channel is the smallest member to be considered. This size is readily available from stock. It is also recommended that a channel larger than 5″ is not used as the panel becomes too wide and unsupported.

Table 2.2 presents the structural characteristics, S_c, section modulus, of various ribs combined with a portion of the panel as a composite section.

For ribs, angles, tubing, and channels not included in Tables 2.2–2.5, calculate the section modulus, S_c, using the Structural Angle – Panel Worksheet (Section A.3), the Structural Tubing – Panel Worksheet (Section A.1), the Structural Channel-Panel Worksheet (Section A.5), or the Reinforcing Rib-Panel Worksheet (Section A.6) in the Appendix A.

The load per inch "w" on the reinforcing member is the reaction of two adjacent panels; therefore, the uniform load is assumed to be twice the reaction: $w = 2R_{max}$.

The total load "L_t" on each panel is

$$P_{Red}ab = 5(15.33 \times 15) = 1149 \text{ lbs} = L_t \tag{2.9}$$

Table 2.2 provides the section modulus, S_c, for rib reinforcing of panels.

Note: The section modulus contributed by the welds is not included due to intermittent welding along the rib. Where welding occurs, there is additional strength.

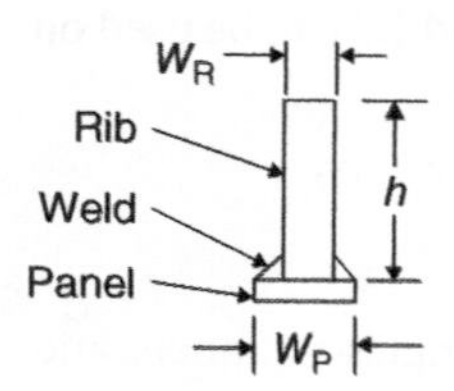

Table 2.2 Rib reinforcing section modulus, S_c.

Rib: $W_R \times h$	**10 ga. (0.1345)**	**8 ga. (0.1644)**	**7 ga. (0.1793)**	**3/16″ (0.188)**	**1/4″ (0.250)**	W_P
1/4 × 2.00	0.275	0.295	0.308	0.318	0.361	0.75
1/4 × 3.00	0.539	0.575	0.591	0.602	0.673	0.75
1/4 × 4.00	0.887	0.933	0.957	1.017	1.070	0.75
1/4 × 5.00	1.311	1.377	1.406	1.425	1.549	0.75
1/4 × 6.00	1.836	1.904	1.939	1.962	2.113	0.75
3/8 × 2.00	0.354	0.396	0.408	0.415	0.468	0.875
3/8 × 3.00	0.745	0.784	0.806	0.813	0.908	0.875
3/8 × 4.00	1.244	1.297	1.323	1.339	1.447	0.875
3/8 × 5.00	1.868	1.937	1.967	1.985	2.125	0.875
3/8 × 6.00	2.617	2.697	2.737	2.758	2.926	0.875
1/2 × 2.00	0.464	0.492	0.507	0.515	0.573	1.00
1/2 × 3.00	0.949	0.992	1.012	1.027	1.114	1.00
1/2 × 4.00	1.600	1.658	1.686	1.703	1.822	1.00

For section modulus values not given in the table, use the "Reinforcing Rib – Panel Worksheet" in the Appendix A.

Table 2.3 provides the section modulus, S_c, for structural angles with panel thicknesses in a composite section.

The section modulus contributed by the welds is not included due to intermittent welding along the angle. Where welding occurs, there is additional strength.

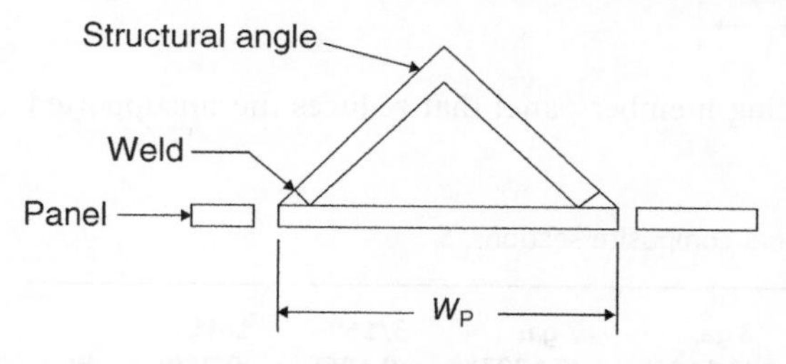

W_p is the width of the reinforcing member-panel that reduces the unsupported panel width.

Table 2.3 Structural angle – panel composite sections.

Member	10 ga. (0.1345)	8 ga. (0.1644)	7 ga. (0.1793)	3/16″ (0.188)	1/4″ (0.250)	W_P
2 × 2 × 1/8	0.770	0.840	0.876	0.892	1.010	3.00
2 × 2 × 3/16	0.900	1.002	1.021	1.086	1.175	3.094
2 × 2 × 1/4	1.027	1.108	1.145	1.164	1.200	3.182
2-1/2 × 2 1/2 × 1/4	1.670	1.805	1.867	1.901	2.133	3.89
3 × 3 × 1/4	2.470	2.684	2.773	2.827	3.190	4.59
3 × 3 × 3/8	3.027	3.228	3.327	3.378	3.755	4.77
4 × 4 × 1/4	4.600	4.983	5.173	5.563	5.966	6.00

For section modulus values not given in the table, use Section A.3.
It is not recommended that an angle larger than 4″ × 4″ is used as the panel under the angle becomes too wide and unsupported.

Table 2.4 provides the section modulus, S_C, for structural tubes with panel thicknesses in a composite section.

Note: The section modulus contributed by the welds is not included due to intermittent welding along the tube. Where welding occurs, there is additional strength.

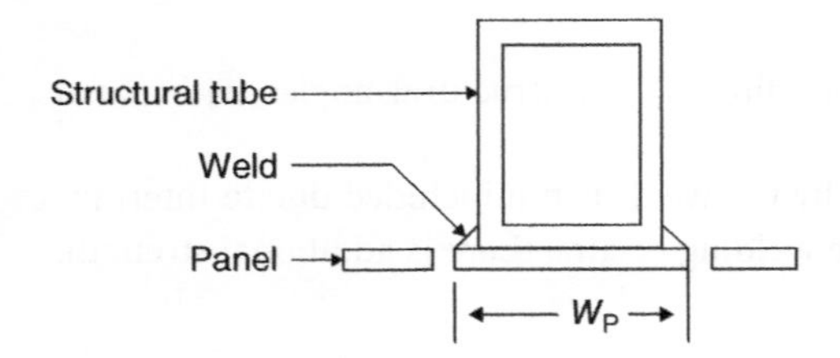

W_P is the width of the reinforcing member-panel that reduces the unsupported panel width.

Table 2.4 Rectangular tube – panel composite sections, S_C.

Member	10 ga. (0.1345)	8 ga. (0.1644)	7 ga. (0.1793)	3/16″ (0.188)	1/4″ (0.250)	W_P
2 × 2 × 3/16	0.996	1.147	1.188	1.214	1.366	2.50
2 × 2 × 1/4	1.185	1.283	1.346	1.366	1.434	2.50
2-1/2 × 2 1/2 × 1/4	1.997	2.157	1.223	2.266	2.584	2.75
3 × 3 × 1/4	2.971	3.150	3.242	3.418	3.653	3.50

Table 2.4 (Continued)

Member	10 ga. (0.1345)	8 ga. (0.1644)	7 ga. (0.1793)	3/16″ (0.188)	1/4″ (0.250)	W_p
4×3×3/16	3.853	4.109	4.242	4.317	4.679	3.50
4×3×1/4	4.406	4.655	4.782	4.852	5.522	3.50
4×3×5/16	4.916	5.096	5.217	5.285	5.772	3.50
4×4×1/4	5.682	6.019	6.179	6.273	6.479	4.50
4×4×3/8	6.797	7.106	7.259	7.349	7.983	4.50
4×4×1/2	7.412	7.703	7.811	7.933	8.196	4.50
6×3×1/4	7.609	7.929	8.195	8.300	9.085	3.50
5×3×1/2	8.054	8.396	8.488	8.533	9.300	3.50
5×4×3/8	9.324	9.734	9.920	10.039	10.840	4.50
6×4×1/4	9.590	10.110	10.360	10.510	11.550	4.50

For section modulus values not given in the table, use Section A.2.

Table 2.5 provides the section modulus, S_c, for channels with panel thicknesses in a composite section.

Note: The section modulus contributed by the welds is not included due to intermittent welding along the channel. Where welding occurs, there is additional strength.

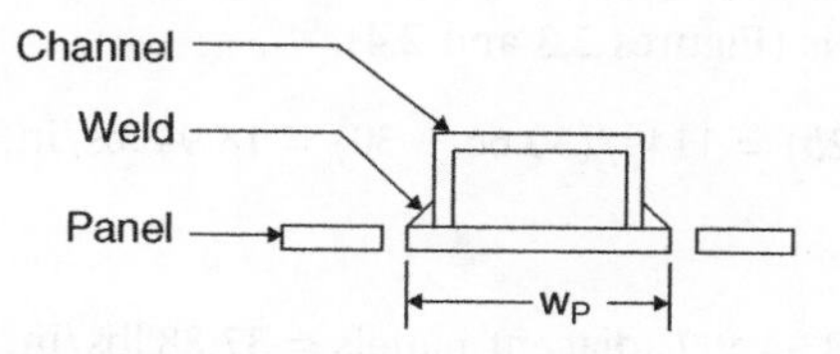

W_p is the width of the reinforcing member-panel that reduces the unsupported panel width.

Table 2.5 Channel – panel composite sections, S_c.

Member	10 ga. (0.1345)	8 ga. (0.1644)	7 ga. (0.1793)	3/16″ (0.188)	1/4″ (0.250)
C3×4.1	0.638	0.712	0.750	0.775	0.909
C3×5	0.711	0.785	0.813	0.844	1.000
C3×6	0.781	0.862	0.904	0.926	1.090

(Continued)

Table 2.5 (Continued)

Member	10 ga. (0.1345)	8 ga. (0.1644)	7 ga. (0.1793)	3/16″ (0.188)	1/4″ (0.250)
C4 × 5.4	0.933	1.042	1.064	1.123	1.328
C4 × 7.25	1.063	1.180	1.236	1.271	1.497
C5 × 6.7	1.273	1.420	1.489	1.535	1.813
C5 × 9	1.429	1.587	1.663	1.709	2.020

For section modulus values not given in the table, use the "Structural Channel-Panel Worksheet" in the Appendix A.
It is not recommended that a channel wider than 5″ is used, as the panel under the channel becomes too wide and unsupported.

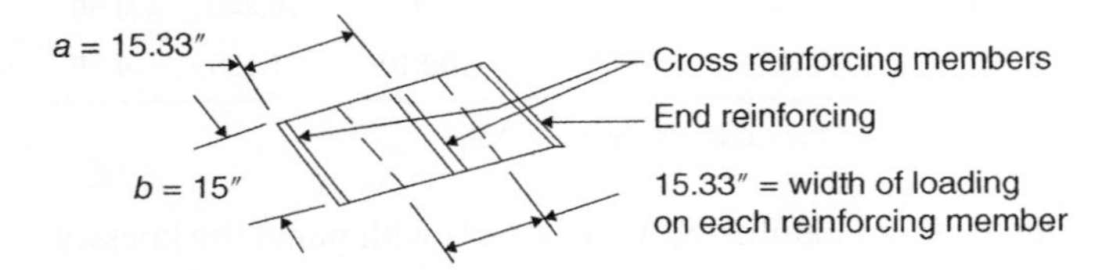

Figure 2.3 Width of loading small panel.

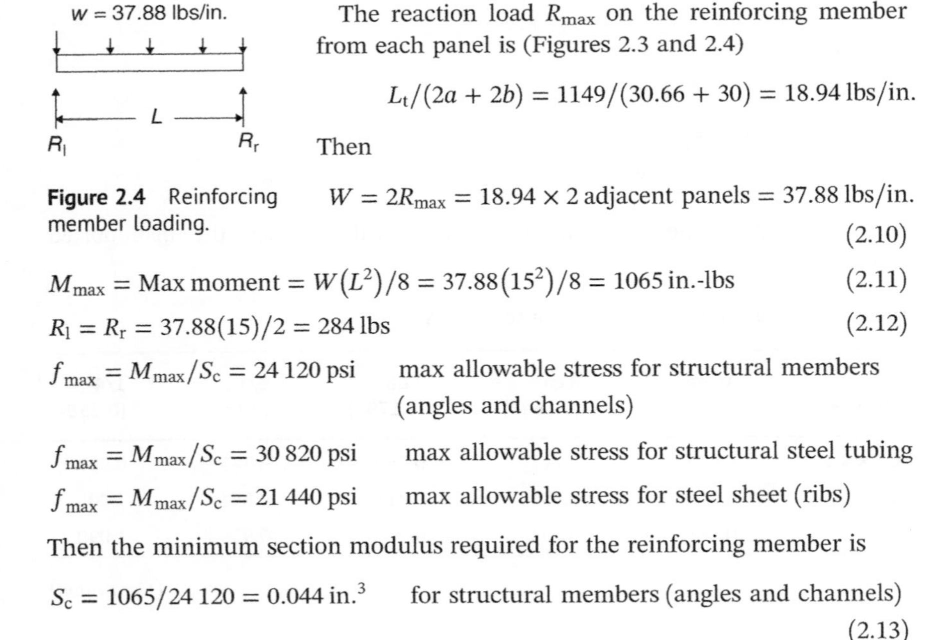

Figure 2.4 Reinforcing member loading.

The reaction load R_{max} on the reinforcing member from each panel is (Figures 2.3 and 2.4)

$$L_t/(2a + 2b) = 1149/(30.66 + 30) = 18.94\ \text{lbs/in.}$$

Then

$$W = 2R_{max} = 18.94 \times 2\ \text{adjacent panels} = 37.88\ \text{lbs/in.} \tag{2.10}$$

$$M_{max} = \text{Max moment} = W(L^2)/8 = 37.88(15^2)/8 = 1065\ \text{in.-lbs} \tag{2.11}$$

$$R_l = R_r = 37.88(15)/2 = 284\ \text{lbs} \tag{2.12}$$

$f_{max} = M_{max}/S_c = 24\,120$ psi max allowable stress for structural members (angles and channels)

$f_{max} = M_{max}/S_c = 30\,820$ psi max allowable stress for structural steel tubing

$f_{max} = M_{max}/S_c = 21\,440$ psi max allowable stress for steel sheet (ribs)

Then the minimum section modulus required for the reinforcing member is

$$S_c = 1065/24\,120 = 0.044\ \text{in.}^3 \quad \text{for structural members (angles and channels)} \tag{2.13}$$

$$S_c = 1065/30\,820 = 0.034 \text{ in.}^3 \quad \text{for structural steel tubing} \tag{2.14}$$

$$S_c = 1065/21\,440 = 0.050 \text{ in.}^3 \quad \text{for steel sheet (rib)} \tag{2.15}$$

From Table 2.2, the smallest rib is 2.0″ × 1/4″ thick with a section modulus of 0.275 in.^{3}. The resultant stress in the rib is then

$$f = 1065/0.275 = 3872 \text{ psi} \tag{2.16}$$

From Table 2.3, under the panel thickness of 10 ga. (0.1345″), the smallest angle is 2″ × 2″ × 1/8″ and has a section modulus S_c of 0.770 in.^{3}

The resultant stress in the cross reinforcing angle is then

$$f = 1065/0.770 = 1383 \text{ psi} \tag{2.17}$$

This is a very low stress.

From Table 2.4, under the panel thickness of 10 ga. (0.1345″), the smallest structural tube is 2″ × 2″ × 3/16″ with a section modulus, $S_c = 0.996$ in.^{3}. It is obvious that the resulting stress in the reinforcing member is very low and can be considered for this application also. It is a matter of choice, although the angle is the more economical section to be used.

From Table 2.5, under the panel thickness of 10 ga. (0.1345″), the smallest channel is C3 × 4.1 with a section modulus, $S_c = 0.638$ in.^{3}.

Example 2.2 Large Panels

The roof of the dust collector is an example of a large 10 ga. (0.1345") panel 38.25″ × 97.75″.

This example provides a method to divide a large panel into smaller panels capable of sustaining the P_{Red} pressure of 5 psig.

$$a/b = 97.75/38.25 = 2.55 \tag{2.18}$$

This value is greater than 2.0; therefore, from Table 2.1, $\beta_1 = 0.500$ and $\beta_2 = 0.25$

$$f_{max} = 0.500(5.0)(38.25^2)/0.1345^2 = 202189 \text{ psi} \tag{2.19}$$

It is obvious that the stresses in the plate would cause catastrophic failure as the stress level exceeds the ultimate strength of the panel (56 000 psi). The wall thickness would have to be $t = 0.413''$ to resist the $P_{Red} = 5.0$ psig.

For comparison, the largest square 10 ga. (0.1345″) panel that will sustain 5 psig is obtained from Table 2.1, for $a/b = 1.0$, $\beta_1 = 0.3078$, and allowable stress = 21 440 psi

$$21\,440 = 0.3078(5.0)(b^2)/0.1345^2 \quad b^2 = 252 \text{ and } b = 15.88'' \tag{2.20}$$

A trial and error approach is to be taken with the roof panel divided into smaller panels. If the 38.25″ width of the panel is divided in half, the panel is then 38.25/2 = 19. 1″. The 97.75″ length of the panel is to be divided into six sections 97.75/6 = 16.29″. The unsupported size of the panels is 3″ smaller than these dimensions when a reinforcing member angle or structural tube is placed and welded to it. *Note*: This is not the case when a rib is the reinforcing member (refer to Figure 2.5). This provides an increase in safety factor and lower stresses in the plate.

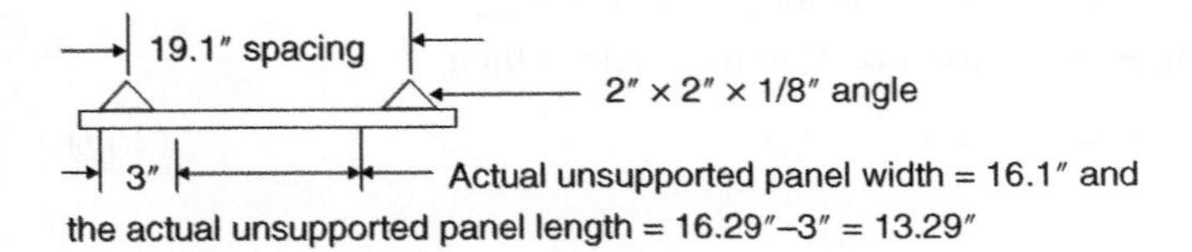

Figure 2.5 Actual unsupported panel width.

$$a/b = 16.1/13.29 = 1.2 \tag{2.21}$$

From Table 2.1, $\beta_1 = 0.3834$ and $\beta_2 = 0.1794$

$$f_{\text{max}} = 0.3834(5.0)(13.29^2)/0.1345^2 = 18\,716 \text{ psi} \quad \text{actual panel stresses} \tag{2.22}$$

The total load "L_t" on each panel is (Figure 2.6)

$$P_{\text{Red}}ab = 5(19.1 \times 16.29) = 1556 \text{ lbs} = L_t \tag{2.23}$$

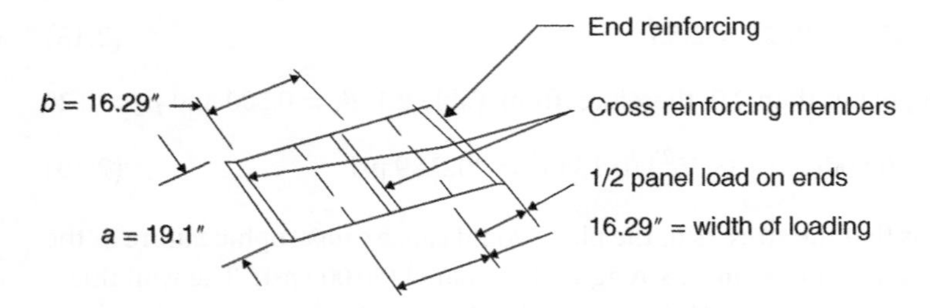

Figure 2.6 Width of loading large panel.

The distributed load "R_{max}" on the cross reinforcing member from each panel is (Figure 2.7)

$$L_t/(2a + 2b) = 1556/(38.2 + 32.58) = 21.98 \text{ lbs/in.} \tag{2.24}$$

Figure 2.7 Cross reinforcing member loading

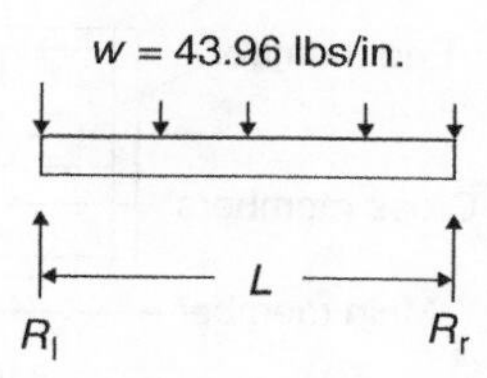

Then

$$W = 2R_{max} = 2 \times 21.98 = 43.96 \text{ lbs/in.} \quad \text{for the two adjacent panels} \tag{2.25}$$

$$M_{max} = \text{max moment} = w(L^2)/8 = 43.96(19.1^2)/8 = 2005 \text{ in.-lbs} \tag{2.26}$$

$$R_l = R_r = 43.96(19.1)/2 = 420 \text{ lbs} \tag{2.27}$$

$$f_{max} = M_{max}/S_c = 24\,120 \text{ psi}$$

is the maximum allowable stress for structural members.

Then the minimum section modulus required for the cross reinforcing member is

$$S_c = 2005/24\,120 = 0.083 \text{ in.}^2 \tag{2.28}$$

From Table 2.2, the rib required would be for a section modulus,

$$S_c = 2005/21\,440 = 0.094 \text{ in.}^3 \tag{2.29}$$

A rib 2.0″ × 1/4″ with a section modulus of 0.275 in.^{3} is adequate.

From Table 2.3, the smallest recommended angle is 2″ × 2″ × 1/8″ and under the 10 ga. (0.1345″) panel column, the section modulus $S_c = 0.770$ in.^{3}.

From Table 2.4, the smallest recommended tube is 2″ × 2″ × 3/16″ under the 10 ga. (0.1345″) panel column with a section modulus $S_c = 0.996$ in.^{3}.

From Table 2.5, the smallest channel is C3–4.1 under the 10 ga. (0.1345″) panel column with a section modulus of 0.638 in.^{3}.

The length of the roof is 97.75 in., and it is decided that a main member will be placed along the entire length of the roof to support the cross members. The end reaction from each of the cross members is (43.96 × 19.1)/2 = 420 lbs. The loads on the main member are 420 × 2 cross reinforcing members = 840 lbs applied every 16.29″ along its length plus the distributed load on the ends from half the panels supported by the end connections. The total load on the member is

$$840 \times 5 = 4200 \text{ lbs} + 43.96 \text{ lbs} \times 16.29/2 = 358 \text{ lbs} \times 2 \text{ end loads} = 716 \text{ lbs} = 4916 \text{ lbs.} \tag{2.30}$$

The total end reactions are (Figure 2.8)

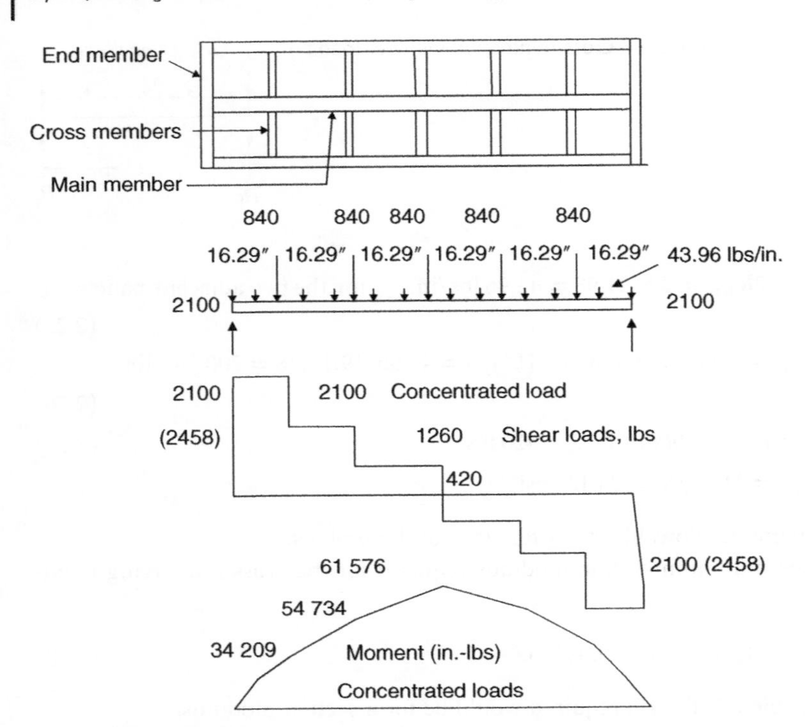

Figure 2.8 Reinforcing member for Example 2.2, shear and moment diagrams.

$$4916/2 = 2458 \text{ lbs} \tag{2.31}$$

By taking the area under the curve in the shear diagram,

$$M_{\text{max}} \text{ for the concentrated loads} = (2100 \times 16.29) + (1260 \times 16.29) + (420 \times 16.29) = 61\,576 \tag{2.32}$$

$$\text{Add the distributed load } M_{\text{max}} = 43.96(97.75^2)/8 = 52\,505 \text{ in.-lbs} + 61\,576 = 11\,4081 \tag{2.33}$$

The required section modulus for an angle,

$$S_c = 114\,081/24\,120 = 4.73 \text{ in.}^3 \tag{2.34}$$

From Table 2.2, the largest rib in the table is 4″ × 1/2″ thick with a section modulus of 1.600 in.^{3}. A reinforcing rib is not practical in this application.

From Table 2.3, the smallest recommended angle under the 10 ga. (0.1345″) panel is 4″ × 4″ × 3/8″ with a section modulus of 5.687 in.^{3}.

For a channel, from Table 2.5, the largest channel under the 10 ga. (0.1345″) panel column is C5–9 with a section modulus of 1.429 in.^{3}. It is obvious that a channel is not practical in this application.

For a structural tube,

$$S_c = 14\,081/30\,820 = 3.701\ \text{in}^3 \tag{2.35}$$

From Table 2.4, the smallest tube section under the 10 ga. (0.1345″) panel column is 4″ × 3″ × 3/16″ with a section modulus of 3.853 in.^{3}.

Dust Collector Side and Rear Panels

Continue with the same reinforcing member size and spacing arrangement that was determined in Example 2.2 for large panels. The side and rear reinforced panel sizes are within the allowable size calculated for large panels in Example 2.2. The access door side of the unit is being reinforced as a separate entity in Chapter 3.

The access door almost completely covers the entire side of the dust collector (Figure 2.9).

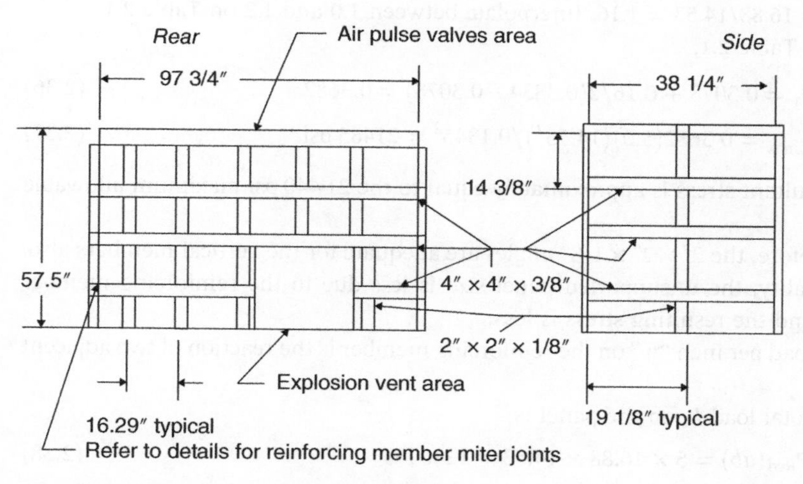

Figure 2.9 Side and rear panel reinforcing.

Hopper Panels: Reinforcing Horizontal and Vertical Members

By inspection of the drawing, Figure 2.1, the panel sizes and reinforcing members determined for the dust collector can be applied to the hopper section (Figure 2.10).

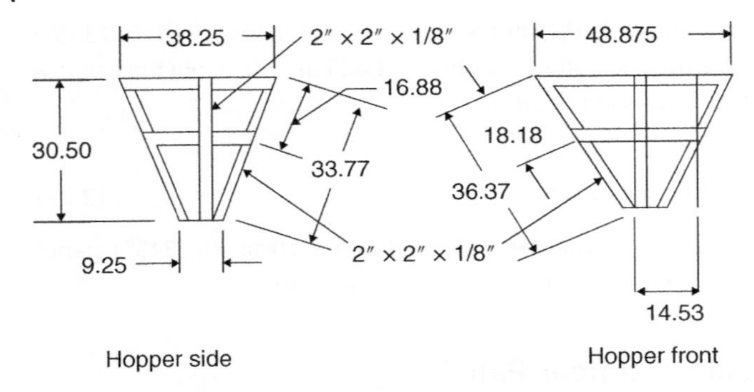

Figure 2.10 Hopper panel reinforcing members.

Divide the side and front hopper panels in half both vertically and horizontally into four panels each. The horizontal reinforcing members are to be 2″ × 2″ × 1/8″ angle. The vertical reinforcing members will be sized for the larger front panels.

Front panels: 33.77/2 × 48.875/2 = 16.88 × 24.44. These panels have a side exceeding 19″. The average width of the panels is (24.44 + 9.25/2) = 14.53″.

a/b = 16.88/14.53 = 1.16. Interpolate between 1.0 and 1.2 on Table 2.1.

From Table 2.1,

$$\beta_1 = 0.3078 + 0.16/2(0.3834 - 0.3078) = 0.3682 \tag{2.36}$$

$$f_{\max} = 0.3682(5.0)\left(14.53^2\right)/0.1345^2 = 21485\ \text{psi} \tag{2.37}$$

The resultant stress is approximately equal to the 21 440 psi maximum allowable stress.

Therefore, the 2″ × 2″ × 1/8″ angles are adequate for the vertical members also. In actuality, the unsupported panel size is less due to the reinforcing member width and the resulting stress is less.

The load per inch "w" on the reinforcing member is the reaction of two adjacent panels.

The total load L_t on the panel is

$$P_{\text{Red}}(ab) = 5 \times 16.88 \times 14.53 = 1226\ \text{lbs} \tag{2.38}$$

39.03 lbs/in.
14.53
R_l R_r

Figure 2.11 The horizontal reinforcing member.

The edge load per inch $R_{\max}$ is

$$L_t/(2a + 2b) = 1226/(33.77 + 29.06) = 19.52 \tag{2.39}$$

Then,

$$w = 2R_{\max} = 2(19.52) = 39.03\ \text{lbs/in.} \tag{2.40}$$

for two panel edges (Figures 2.11 and 2.12).

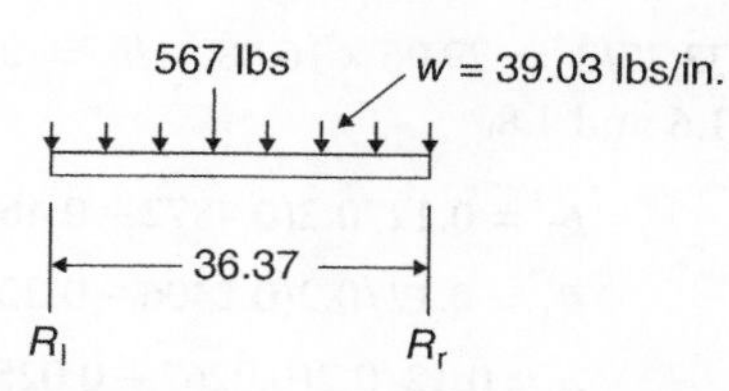

Figure 2.12 Vertical reinforcing member (conservative analysis).

$$R_l = R_r = 39.03(14.53)/2 = 284 \text{ lbs} \quad \text{Concentrated load, } L_c = 2R = 567 \text{ lbs} \tag{2.41}$$

$$\text{Distributed load}: M_{max} = wL^2/8 = 39.03(36.37^2/8) = 6433 \text{ in.-lbs} \tag{2.42}$$

$$\text{Concentrated load}: M_{max} = L_c L/4 = 567 \times 36.37/4 = 5156 \text{ in.-lbs} \tag{2.43}$$

$$\text{Total}: M_{max} = 6453 + 5156 = 11\,609 \text{ in.-lbs} \tag{2.44}$$

The required section modulus,

$$S_c = 11\,609/24\,120 = 0.481 \text{ in.}^3. \tag{2.45}$$

From Table 2.3, for the 2″ × 2″ × 1/8″ angle the section modulus is 0.770 in.^{3}. This angle is adequate for the main vertical member also. The reinforcing members are lightly stressed; therefore, the simplified analysis for these hopper members is adequate. There is a safety factor

$$0.770/0.481 = 1.60 \tag{2.46}$$

Hopper Panels: Using Horizontal Members Only

The rectangular panel stresses are obtained using the formulas for "f_{max}," where the panel has all edges fixed (Figure 2.13). From Table 2.1, (48.875 + 9.25)/2 ×

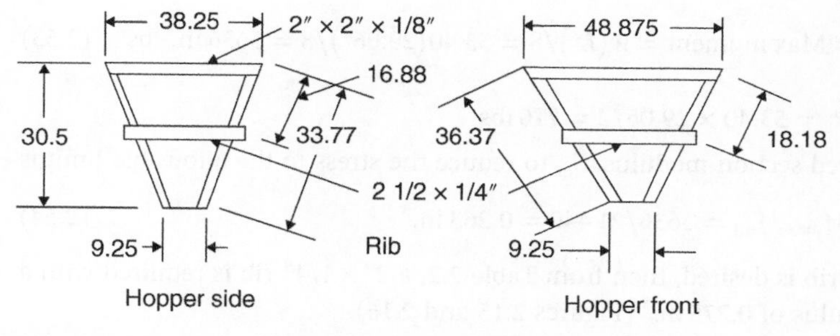

Figure 2.13 Hopper panel horizontal members only.

33.77/2 = 29.06 × 16.88, a/b = 29.06/16.88 = 1.72 and interpolating between 1.6 and 1.8.

$$\beta_1 = 0.12/0.2(0.4872 - 0.4680) + 0.4680 = 0.4795 \tag{2.47}$$

$$\beta_2 = 0.12/0.2(0.2406 - 0.2286) + 0.2286 = 0.2358 \tag{2.48}$$

$$\alpha = 0.12/0.2(0.0267 - 0.0251) + 0.0251 = 0.0261 \tag{2.49}$$

Solving for the allowable pressure, P_{Red}, using the allowable stress of 21 440 psi for steel sheet (2/3 of the 0.2% yield strength):

The sheet thickness is 10 ga. (0.1345″).

The higher stress is at the center of the long edge; therefore, factor β_1 is controlling.

$$P_{\text{Red}}\ \text{allowable} = f_{\max} t_{\text{p}}^2/\beta_1 b^2 = (21\,440)(0.1345^2)/0.4795(12.22^2) = 5.42\ \text{psi} \tag{2.50}$$

The spacing of the reinforcing is adequate to allow the panels to sustain a P_{Red} of 5.0 psig as per the examples being used in this analysis.

The total load T_{t} on a panel is

$$P_{\text{Red}}ab = 5.0 \times 29.06\ \text{avg} \times 16.88 = 2453\ \text{lbs} \tag{2.51}$$

The load per inch of panel edge is

$$T_{\text{t}}/(2a + 2b) = 2453/(58.12 + 33.76) = 26.69\ \text{lbs/in.}$$

The load on each rib is from two panel edges (Figure 2.14):

$$26.69 \times 2 = 53.40\ \text{lbs/in.} \tag{2.52}$$

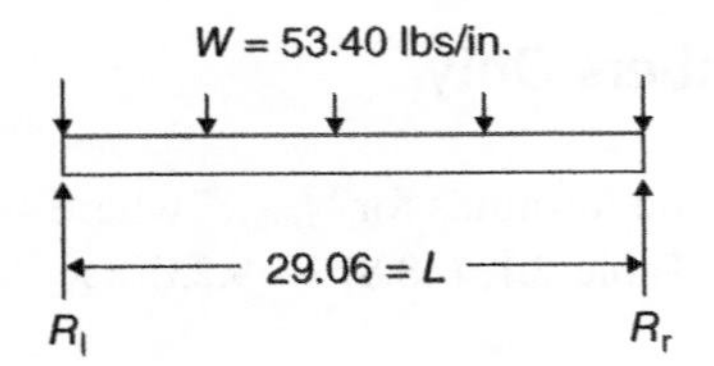

Figure 2.14 Panel reinforcing rib loading.

$$M_{\max} = \text{Max moment} = w(L^2)/8 = 53.40(29.06^2)/8 = 5636\ \text{in.-lbs} \tag{2.53}$$

$$R_{\text{l}} = R_{\text{r}} = 53.40 \times 29.06/2 = 776\ \text{lbs}$$

The required section modulus, S_{c}, to reduce the stress to the allowable limit is

$$S_{\text{c}} = M_{\max}/f_{\text{all}} = 5636/21\,440 = 0.263\ \text{in.}^3 \tag{2.54}$$

If a single rib is desired, then from Table 2.2, a 2″ × 1/4″ rib is required with a section modulus of 0.275 in.^{3} (Figures 2.15 and 2.16).

Figure 2.15 Panel reinforcing rib.

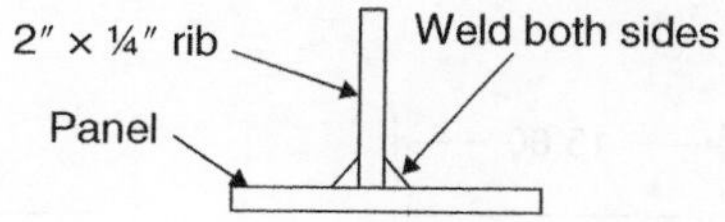

2″ × ¼″ rib
Weld both sides
Panel

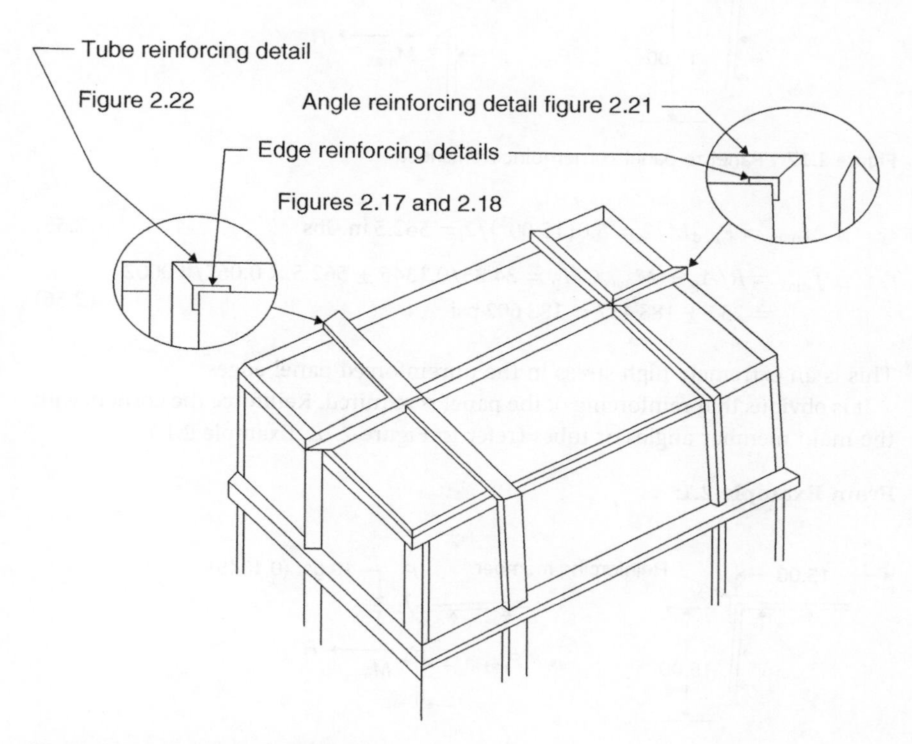

Figure 2.16 Square/rectangular vessel.

From Example 2.1:
To simplify the analysis, assume $W_P = a\ 1''$ strip beam with fixed ends. Panel corner edges unreinforced (Figure 2.17).

$$P_{\text{Red}} = 5.0\ \text{psi}, R = 34.35\ \text{lbs/in.}$$

$$A_p = W_p(t_p) = 1''(0.1345) = 0.1345\ \text{in.}^2$$

$$I_p = W_p\left(t_p^3\right)/12 = 1'' \times \left(0.1345^3\right)/12 = 0.0002\ \text{in.}^4$$

$$C = t_p/2 = 0.1345/2 = 0.067$$

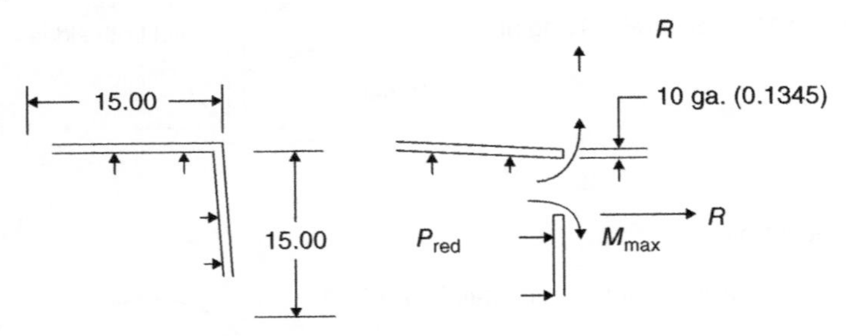

Figure 2.17 Panel to panel corner joint worksheet.

$$M_{max} = P_{Red}L^2/2 = 5.0(15.00^2)/2 = 562.5 \text{ in.-lbs} \tag{2.55}$$

$$\begin{aligned} f_{max} &= R/A_p \pm M_{max}C/I_p = 34.35/0.1345 \pm 562.5 \times 0.067/0.0002 \\ &= 255 \pm 188\,437 = 188\,602 \text{ psi} \end{aligned} \tag{2.56}$$

This is an extremely high stress in the unreinforced panel edges.

It is obvious that reinforcing of the panel is required. Reinforce the corners with the main member angles or tubes (refer to Figure 2.18, Example 2.1.)

From Example 2.1:

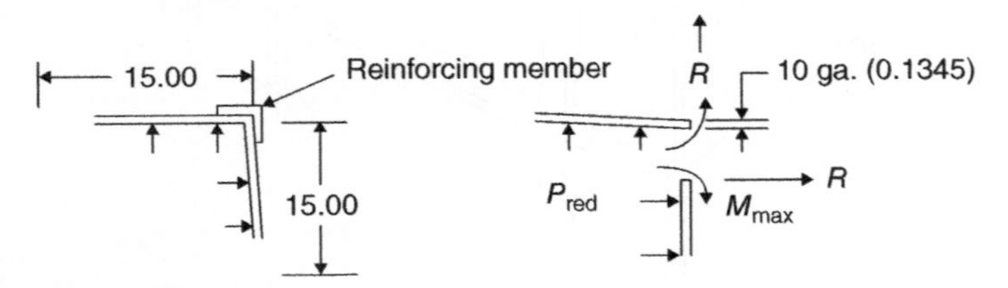

Figure 2.18 Panel to panel corner joint reinforcing worksheet.

To simplify the analysis, assume $W_p = 1''$ strip beam with fixed ends. Panel corner edges reinforced with $4'' \times 4'' \times 3/8''$ angle (the same as the main member reinforcing).

$$P_{Red} = 5.0 \text{ psi}, R = 34.35 \text{ lbs/in.}, A_p = W_p(t_p + t_r) = 1(0.1345 + 0.375) = 0.510 \text{ in.}^2 \tag{2.57}$$

$$I_p = W_p(t_p + t_r)^3/12 = 1(0.1345 + 0.375)^3/12 = 0.011 \text{ in.}^4 \tag{2.58}$$

$$C = (t_p + t_r)/2 = 0.510/2 = 0.255'' \tag{2.59}$$

$$M_{max} = P_{Red}L^2/2 = 5.0 \times 15.00^2/2 = 562.5 \text{ in.-lbs} \tag{2.60}$$

$$\begin{aligned} f_{max} &= R/A_p \pm M_{max}C/I_p = 34.35/0.510 \pm 562.5 \times 0.255/0.011 \\ &= 67.35 \pm 13\,040 = 13\,107 \text{ psi} \end{aligned} \tag{2.61}$$

$$\text{Factor of safety} = 21\,440/13\,107 = 1.64 \tag{2.62}$$

Reinforce the corners with the main member angles or tube to provide a safe structure.

From Example 2.2:

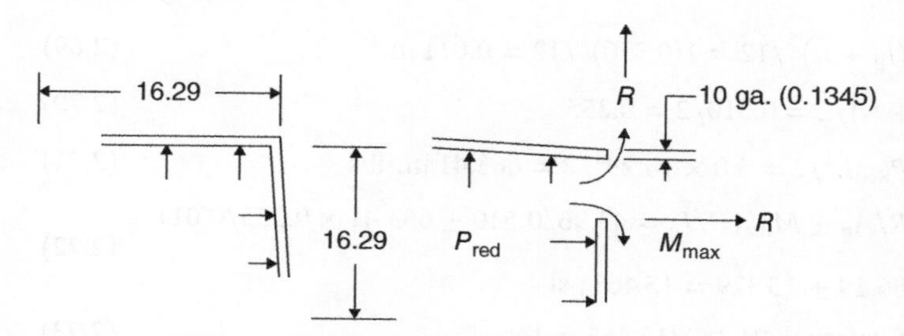

Figure 2.19 Panel to panel corner joint worksheet.

To simplify the analysis, assume $W_P = a$ 1″ strip beam with fixed ends. Panel corner edges unreinforced (Figure 2.19).

$$P_{Red} = 5.0\,\text{psi}, R = 43.96\,\text{lbs/in.}, A_p = W_p t_p = 1 \times 0.1345 = 0.1345\,\text{in.}^2 \tag{2.63}$$

$$I_p = W_p t_p^3/12 = 1 \times 0.1345^3/12 = 0.0002\,\text{in.}^4 \tag{2.64}$$

$$C = t_p/2 = 0.1345/2 = 0.067 \tag{2.65}$$

$$M_{max} = P_{Red}L^2/2 = 5.0 \times 16.29^2/2 = 663.41\,\text{in.-lbs} \tag{2.66}$$

$$\begin{aligned} f_{max} &= R/A_p \pm M_{max}C/I_p = 43.46/0.1345 \pm 663.41 \times 0.067/0.0002 \\ &= 323 \pm 188\,435 = 222\,242\,\text{psi} \end{aligned} \tag{2.67}$$

This is an extremely high stress in the unreinforced panel edges.

It is obvious that reinforcing of the panel is required. Reinforce the corners with the main member angles or tubes. Refer to Figure 2.20, Example 2.2.

From Example 2.2:

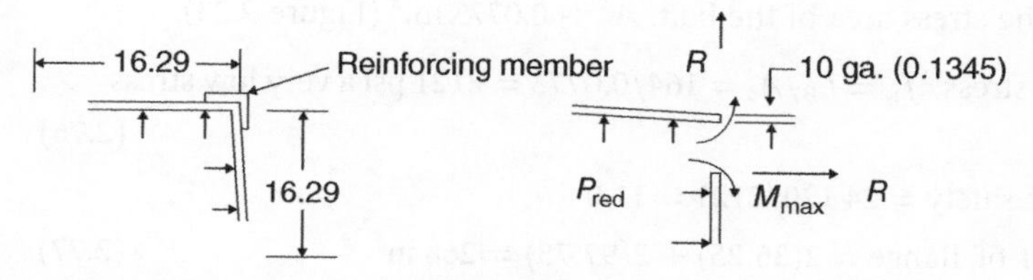

Figure 2.20 Panel to panel corner joint reinforcing worksheet.

To simplify the analysis, assume $W_p = 1''$ strip beam with fixed ends. Panel corner edges reinforced with $4'' \times 4'' \times 3/8''$ angle (the same as the main member reinforcing).

$$P_{Red} = 5.0 \text{ psi}, R = 43.96 \text{ lbs/in.}, A_p = W_p(t_p + t_r) = 1(0.1345 + 0.375) = 0.510 \text{ in.}^2 \quad (2.68)$$

$$I_p = W_p(t_p + t_r)^3/12 = 1(0.510)^3/12 = 0.011 \text{ in.}^4 \quad (2.69)$$

$$C = (t_p + t_r)/2 = 0.510/2 = 0.255 \quad (2.70)$$

$$M_{max} = P_{Red}L^2/2 = 5.0 \times 16.29^2/2 = 663.41 \text{ in.-lbs} \quad (2.71)$$

$$f_{max} = R/A_p \pm M_{max}C/I_p = 43.96/0.510 \pm 663.41 \times 0.255/0.011 = 86.19 \pm 15\,379 = 15\,465 \text{ psi} \quad (2.72)$$

$$\text{Factor of safety} = 21\,440/15\,465 = 1.4 \quad (2.73)$$

Reinforce the corners with the main member angles or tube to provide a safe structure (Figures 2.21 and 2.22).

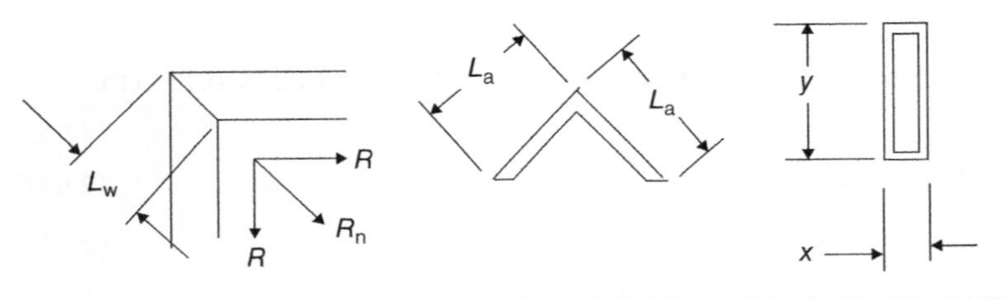

Figure 2.21 Reinforcing cross member miter joint.

The total load, L_T, on the flange $= P_{Red} \times 36.25 \times 97.75 = 5 \times 3543 = 17\,717$ lbs (2.74)

$$\text{The load per bolt, } L_B = L_T/N_B = 17\,717/108 = 164 \text{ lbs} \quad (2.75)$$

From Figures 4.7 and 4.8, the allowable 67% yield strength for mild steel bolts is 24 120 psi and the stress area of the bolt, A_s, is 0.0773 in.^{2} (Figure 2.23).

$$\text{The bolt stress, } f_b = L_B/A_s = 164/0.0773 = 2121 \text{ psi a very low stress} \quad (2.76)$$

$$\text{Factor of safety} = 24\,120/2121 = 11.0$$

$$\text{Perimeter of flange} = 2(36.25) + 2(97.75) = 268 \text{ in.} \quad (2.77)$$

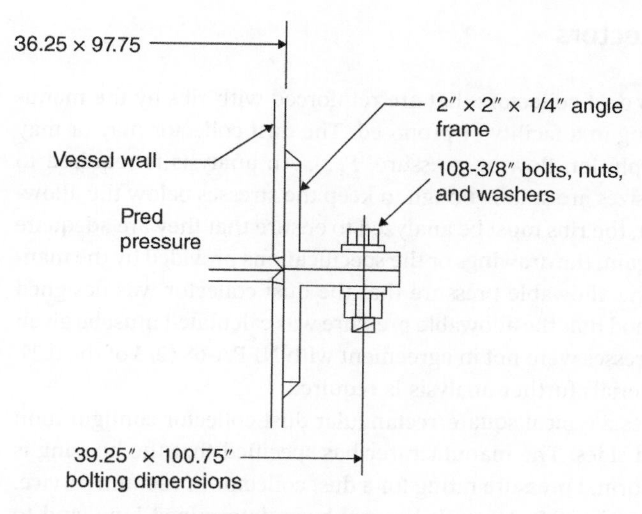

Figure 2.22 Bolted flange detail.

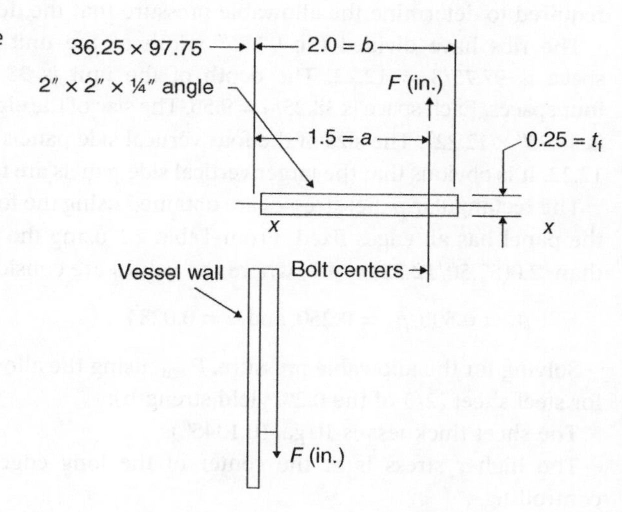

Figure 2.23 Bolted flange stress.

$$F\text{ (in.)} = L_T/\text{perimeter of flange} = 17\,717/268 = 66.1\text{ lbs/in.}$$

$$I_{x-x} = bt_f^3/12 = 2.0(0.25^3)/12 = 0.0026\text{ in.}^4 \quad (2.78)$$

$$M_t = F\text{ (in.)} \times a = 66.1 \times 1.5 = 99.15\text{ in.-lbs/in.} \quad (2.79)$$

$$C = t_f/2 = 0.25/2 = 0.125\text{ in.}$$

$$\text{Stress} = M_t C/I_{x\text{-}x} = 99.15 \times (0.125)/0.0026 = 4766\text{ psi} \quad (2.80)$$

$$\text{Factor of safety} = 24\,120/4766 = 5.0 \quad (2.81)$$

Ribbed Dust Collectors

This section applies to dust collectors that are reinforced with ribs by the manufacturer and are existing in a facility or proposed. The dust collector may or may not be safe for the explosion flowing pressure, P_{Red}. An analysis is required to ensure that the panel sizes are small enough to keep the stresses below the allowable limits. In addition, the ribs must be analyzed to ensure that they are adequate to sustain the loads. Again, the drawings or the specifications provided by the manufacturer must state the allowable pressure that the dust collector was designed and built for. The method that the allowable pressure was calculated must be given and if the allowable stresses were not in agreement with NFPA-68 (2/3 of the 0.2% yield stress of the material) further analysis is required.

Figure 2.24 illustrates a typical square/rectangular dust collector configuration reinforced with ribbed sides. The manufacturer has specified that the housing is rated for ±20″ w.g. a normal pressure rating for a dust collector in vacuum service. The equivalent pressure is ±0.723 psig. It has not been determined how, and to what standard, the pressure rating was calculated; therefore, an analysis is required to determine the allowable pressure that the dust collector can sustain.

The ribs have divided the 97.75″ width of the unit into eight spaces. Each space is 97.75/8 = 12.22. The depth of the unit is 38.25″ and is divided into four spaces. Each space is 38.25/4 = 9.56. The size of the eight horizontal roof panels is 38.25″ × 12.22″. The size of the four vertical side panels is 136 − 78.50 = 57.50 × 12.22. It is obvious that the larger vertical side panels are to be analyzed for stress.

The rectangular panel stresses are obtained using the formulas for "f_{max}," where the panel has all edges fixed. From Table 2.1 using the values for a/b = greater than 2.0(57.50/12.22) = 4.7, where the values are considered infinite:

$$\beta_1 = 0.500, \beta_2 = 0.250, \text{and } \alpha = 0.0284$$

Solving for the allowable pressure, P_{Red}, using the allowable stress of 21 440 psi for steel sheet (2/3 of the 0.2% yield strength):

The sheet thickness is 10 ga. (0.1345″).

The higher stress is at the center of the long edge; therefore, factor β_1 is controlling.

$$P_{Red}\text{ allowable} = f_{max}t_p^2/\beta_1 b^2 = (21\,440)(0.1345^2)/(0.500)(12.22^2) = 5.2\text{ psi} \quad (2.82)$$

The spacing of the ribs is adequate to allow the panels to sustain a P_{Red} of 5.0 psig as per the examples being used in this analysis. The ribs must be analyzed to determine whether or not they are adequate to sustain the 5.0 psig loading.

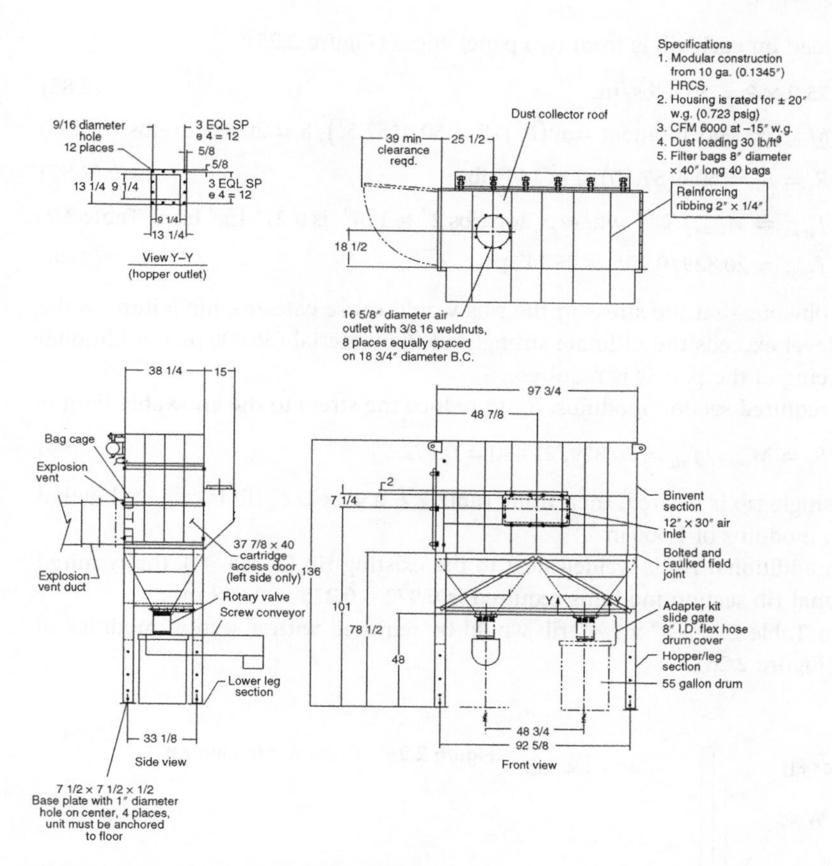

Figure 2.24 Typical dust collector ribbed.

The total load T_t on a panel is

$$P_{Red} \times ab = 5.0 \times 57.50 \times 12.22 = 3513 \text{ lbs} \tag{2.83}$$

The load per inch of panel edge is

$$T_t/(2a + 2b) = 3513/(115 + 24.44) = 25.2 \text{ lbs/in.} \tag{2.84}$$

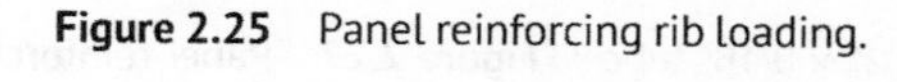
Figure 2.25 Panel reinforcing rib loading.

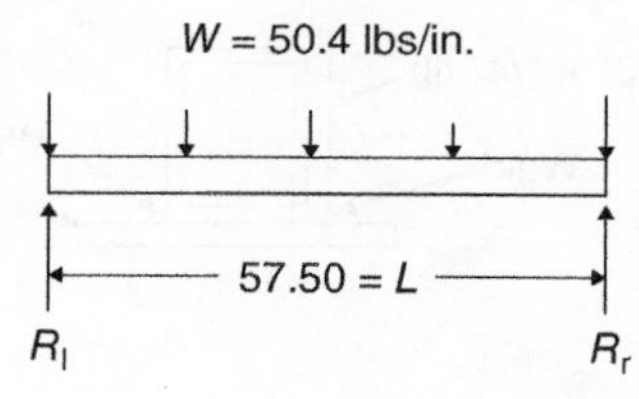

The load on each rib is from two panel edges (Figure 2.25):

$$25.2 \times 2 = 50.4\ \text{lbs/in.} \tag{2.85}$$

$$M_{max} = \text{Max moment} = w(L^2)/8 = 50.4(57.5^2)/8 = 20\,829\ \text{in.-lbs} \tag{2.86}$$

$$R_l = R_r = 50.4(57.50)/2 = 1449\ \text{lbs} \tag{2.87}$$

$f_{max} = M_{max}/S_c$ where S_c for ribs 2″ × 1/4″ is 0.275 in.^{3} from Table 2.2

$$f_{max} = 20\,829/0.275 = 75\,742\ \text{psi} \tag{2.88}$$

It is obvious that the stress in the ribs would cause catastrophic failure as the stress level exceeds the ultimate strength of the material (56 000 psi). Additional reinforcing of the panels is required.

The required section modulus, S_c, to reduce the stress to the allowable limit is

$$S_c = M_{max}/f_{all} = 20\,829/21\,440 = 0.972 \tag{2.89}$$

If a single rib is desired, then from Table 2.2, a 4″ × 1/2″ rib is required with a section modulus of 1.600 in.^{3}.

If an additional rib is welded next to the existing rib (2″ × 1/4″), the required additional rib section modulus required is 0.972 − 0.275 = 0.698 in.^{3}.

From Table 2.2, a 4″ × 1/4″ rib would be required with a section modulus of 0.897 (Figure 2.26).

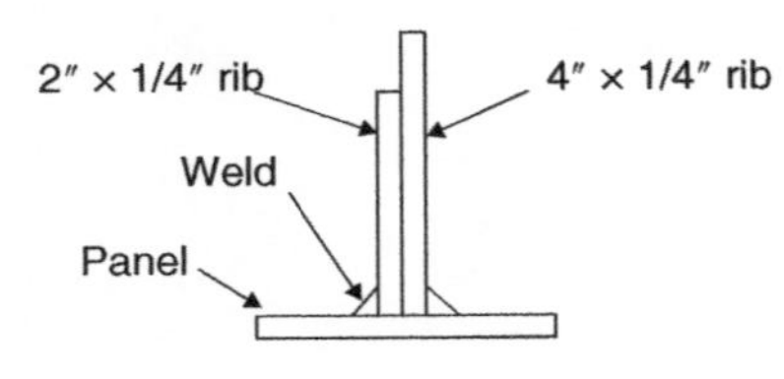

Figure 2.26 Panel reinforcing rib.

Another method of additional reinforcing is to weld a rectangular tube next to the existing ribs: the structural tube has a higher allowable stress (30 820 psi) than a sheet rib; therefore, the required tube section modulus is then 0.806 × 21 440/30 820 = 0.561 in.^{3}. From Table 2.4, under the 10 ga. Panel column, a 2″ × 2″ × 3/16″ tube (the smallest recommended size) with a section modulus of 0.996 in.^{3} is more than adequate (Figure 2.27).

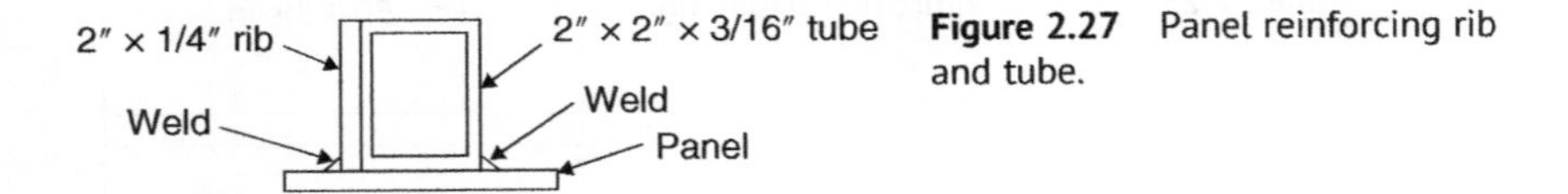

Figure 2.27 Panel reinforcing rib and tube.

3

Round/Cylindrical Dust Collectors

The cylinder and the flat head wall thicknesses are 10 ga. (0.1345″) for this example (Figure 3.1).

A cylindrical dust collector is usually very structurally sound except for the roof, if it is a flat head and has any access doors or ports.

A flat head roof is very stiff and causes higher stresses in the cylinder than a hemispherical or ellipsoidal head. The moment and shear at the fixed end of the cylinder under the action of internal pressure, P_{Red}, must be analyzed to determine the discontinuity stresses of the joint (Figure 3.2).

$P_{\mathrm{Red}} = 5.0$ psig (assumed consistent with continuing analysis)
t_{h} = Head thickness, in.
t_{c} = Cylinder thickness, in.
R_{o} = Cylinder radius, in. (assume 25″)
E = Modulus of elasticity, 30×10^6 psi
$\nu = 0.3$

The growth of a free cylinder under the action of a uniform internal pressure is

$$\Delta = (1 - \nu/2) P_{\mathrm{Red}} {R_{\mathrm{o}}}^2 / E t_{\mathrm{c}} \tag{3.1}$$

$$\Delta = (1 - 0.3/2) 5.0 (0.25^2) / 30 \times 10^6 (0.1345) = 0.000\,66 \text{ in.} \tag{3.2}$$

The growth and slope at the fixed end must be equal to 0, the moment and shear are determined from the conditions:

$$\Delta = 1/2\beta^3 D(\beta M_{\mathrm{o}} + V_{\mathrm{o}}) \tag{3.3}$$

$$\Delta' = 1/2\beta^2 D(2\beta M + V_{\mathrm{o}}) \tag{3.4}$$

Where in this case: $\Delta = 0.000\,66$

Hence,

$$\beta M_{\mathrm{o}} + V_{\mathrm{o}} = 1/2\beta^3 D = 0.000\,66 \tag{3.5}$$

Explosion Vented Equipment System Protection Guide, First Edition. Robert C. Comer.

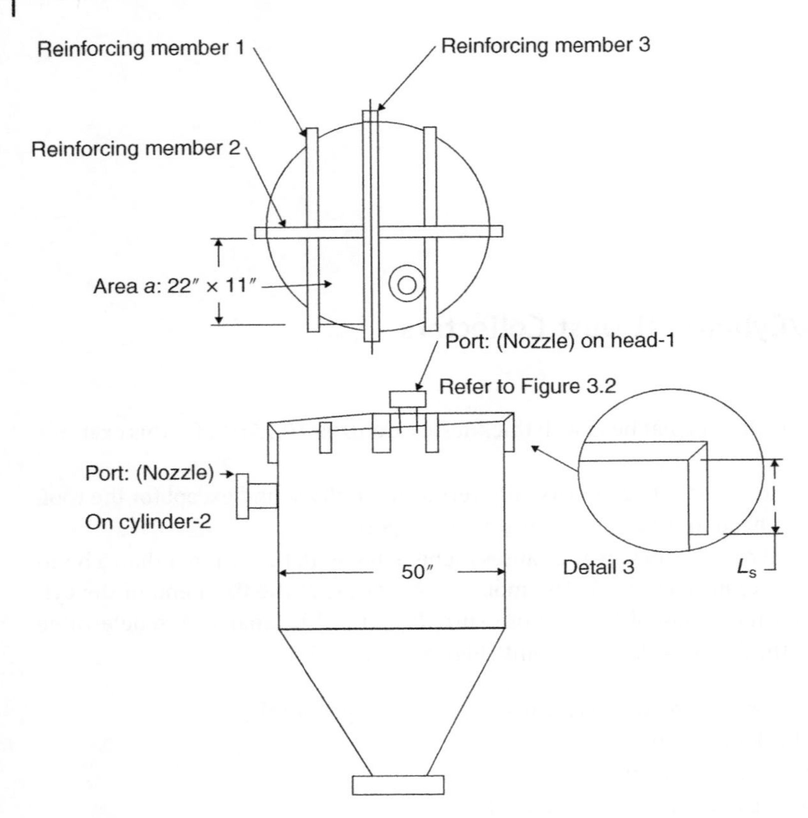

Figure 3.1 Cylindrical vessel.

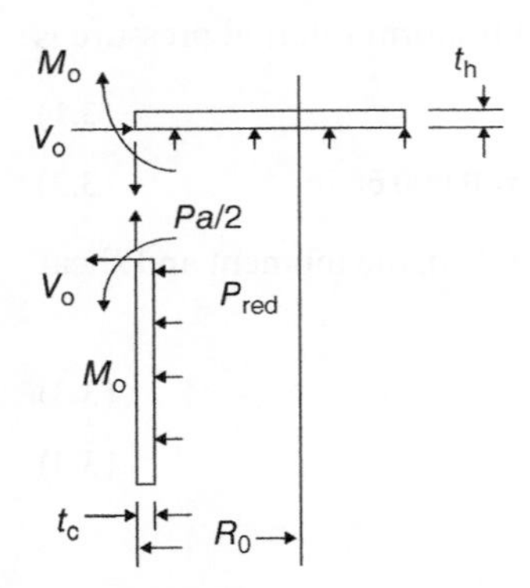

Figure 3.2 Cylinder and head stresses.

$$2\beta M_o + V_o = 0 \tag{3.6}$$

$$M_o = 0.000\,66 \times 2\beta^2 D \tag{3.7}$$

$$D = Et_c^{\,3}/12(1-\nu^2) = 30 \times 10^6(0.1345^3)/12(1-0.3^2) = 6684 \tag{3.8}$$

$$\beta^2 = \left[3(1-\nu^2)/R^2 t_c^{\,2}\right]^{1/2} = \left[3(1-0.3^2)/25^2(0.1345^2)\right]^{1/2} = 0.491 \tag{3.9}$$

$$M_o = 0.000\,66 \times 2(0.491)(6684) = 4.33 \text{ in.-lbs/in.} \tag{3.10}$$

$$V_o = -2\beta M_o = -2\sqrt{0.491}(4.33) = 6.068 \text{ lbs/in.} \tag{3.11}$$

$$\sigma_x = P_{Red}R_o \pm 6M_o/t_c^{\,2} = 5.0 \times 0.25 \pm 6(4.33)/(0.1345^2)$$
$$= +1561 \text{ psi and} -1311 \text{ psi} \tag{3.12}$$

$$T = V_o/t_c = 6.068/0.1345 = 45.1 \text{ psi} \tag{3.13}$$

$$\sigma_y = P_{Red}R_o/2 \pm 6\nu M_o/t_c^{\,2} = 5.0 \times 25/2 \pm 6(0.3)4.33/(0.1345^2)$$
$$= +493 \text{ psi and} -368 \text{ psi} \tag{3.14}$$

The cylinder is very safe.

The flat head stresses:

Assume a simply supported flat plate with uniform loading, edge moment, and edge normal load.

$$\sigma_{max} = (\sigma_r)_{r=0} = (\sigma_\vartheta)_{r=0} \tag{3.15}$$

$$\sigma_{max} = V_o/t_h \pm 6M_o/t_h^{\,2} \mp 3(3+\nu)R_o^{\,2}/8t_h^{\,2} \times P_{Red} \tag{3.16}$$

Upper and lower signs refer to inner and outer surfaces, respectively.

$$\sigma_{max} = -6.068/0.1345 \pm 6(4.33)/(0.1345^2) \mp 3(3+0.3)(25^2)/8(0.1345^2) \times 5.0 \tag{3.17}$$

$$\sigma_{max} = -45.1 \pm 1436 \mp 213\,771 = +212\,285 \text{ psi and} -215\,257 \text{ psi} \tag{3.18}$$

It is obvious that the head overstresses to the point of a catastrophic failure. The head must be reinforced.

Head (roof) reinforcing member 1 (Figure 3.3):

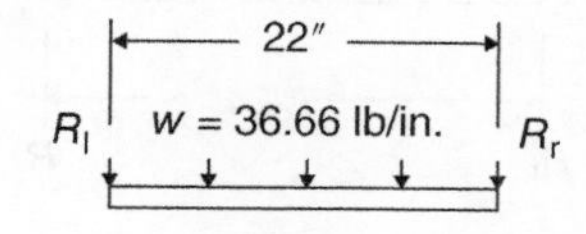

Figure 3.3 Reinforcing member 1 loading.

Area $a = 22 \times 11 = 242$ in.^{2}

Total load on area a P_{Red} = (area a) = $5.0 \times 242 = 1210$ lbs (3.19)

Perimeter of area $a = (2 \times 22) + (2 \times 11) = 66$ in. (3.20)

The load per inch of perimeter $a = 1210/66 = 18.33$ lbs/in. (3.21)

Assume that two adjacent panels load reinforcing member 1 = 18.33 $\times$ 2 = 36.66 lbs-in.

$$M_{max} = \text{Maximum moment} = wL^2/8 = 36.66\left(22^2\right)/8 = 2218 \text{ in.-lbs} \tag{3.22}$$

The end reactions are $R_l = R_r = 2218/2 = 1109$ lbs (3.23)

$$\text{The required section modulus, } S_x = M_{max}/S_{all} = 2218/21\,440 = 0.103 \text{ in.}^3 \tag{3.24}$$

From Table 2.2, a 2″ × 1/4″ thick rib ($S_c = 0.275$ in.^{3}) is adequate for reinforcing member 1 (Figure 3.4).

$$\begin{aligned} M_{max} &= wL^2/8 + RL/4 = 36.66\left(24^2\right)/8 + 1109(24)/4 = 21\,116 + 6654 \\ &= 27\,770 \text{ in.-lbs} \end{aligned} \tag{3.25}$$

The end reactions are $[(36.66 \times 24) + 1109]/2 = 994.4$ lbs (3.26)

$$\text{The required section modulus, } S_x = M_{max}/S_{all} = 27\,770/21\,440 = 1.30 \text{ in.}^3 \tag{3.27}$$

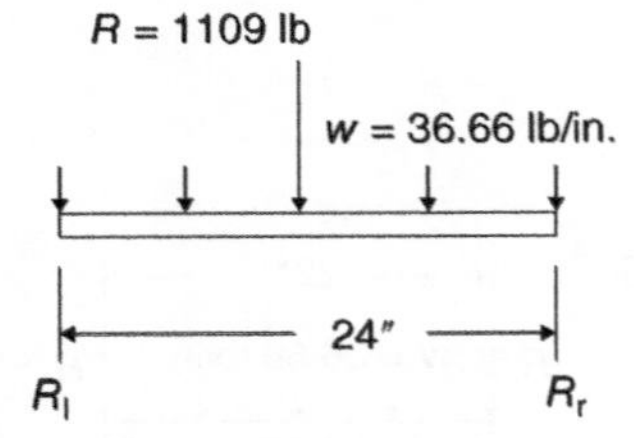

Figure 3.4 Reinforcing member 2 loading.

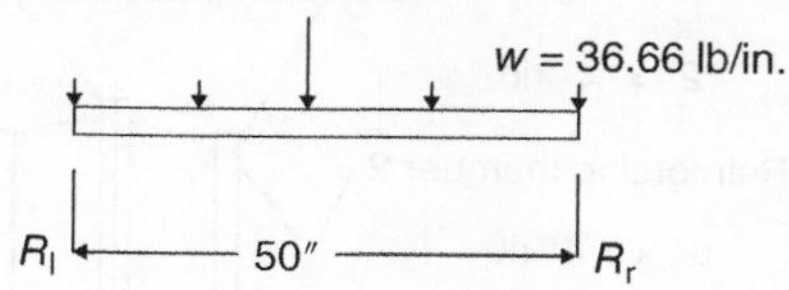

Figure 3.5 Reinforcing member 3 loading.

From Table 2.2, a 5″ × 3/8″ rib ($S_c = 1.868$ in.^{3}) is adequate (Figure 3.5).

$$M_{max} = wL^2/8 + RL/4 = 36.66(50^2)/8 + 2218(50)/4$$
$$= 11\,456 + 27\,725 = 39\,181 \text{ in.-lbs} \quad (3.28)$$

The end reactions are $= (36.66 \times 50 + 2218)/2 = 4051$ lbs

The required section modulus $= S_x = M_{max}/S_{all}$

$$= 38\,181/21\,440 = 1.83 \text{ in.}^3 \quad (3.29)$$

From Table 2.2, a 6″ × 1/4″ rib ($S_c = 1.836$ in.^{3}) is adequate.

For a structural member, the required $S_c = 38\,181/24\,120 = 1.624$ in.^{3} (3.30)

From Table 2.3, under the 10 ga. (0.1345″) column, a 2-1/2″ × 2-1/2″ × 1/4″ angle ($S_c = 1.67$ in.^{3}) is adequate.

From Table 2.5 under the 10 ga. (0.1345″) column, a channel is not practical to be used in this application.

For a structural tube, the required section modulus $= 39\,181/30\,820 = 1.271$ in.^{3} (3.31)

From Table 2.4, under the 10 ga. (0.1345″) column, a 2-1/2″ × 2-1/2″ × 1/4″ (S_c = 1.997 in.^{3}) is adequate (Figure 3.6).

Reinforcing member edge supports (Figures 3.7 and 3.8):

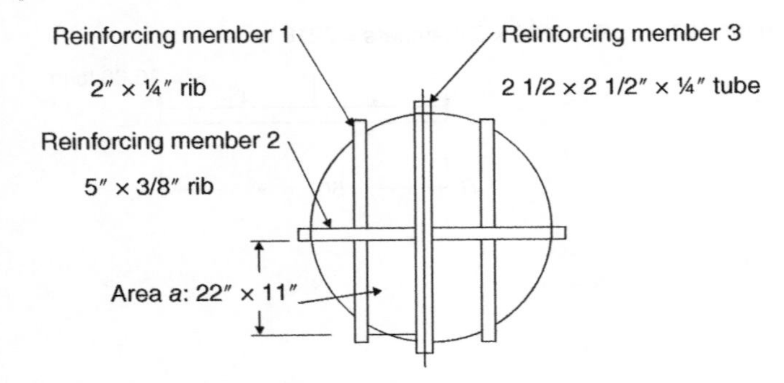

Figure 3.6 Reinforcing members summary.

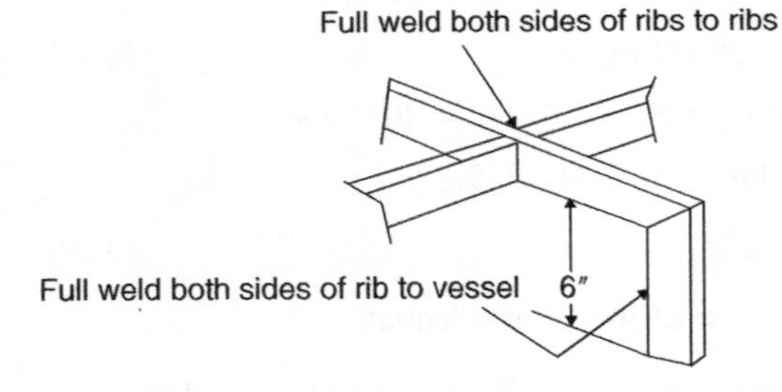

Figure 3.7 Rib reinforcing edges.

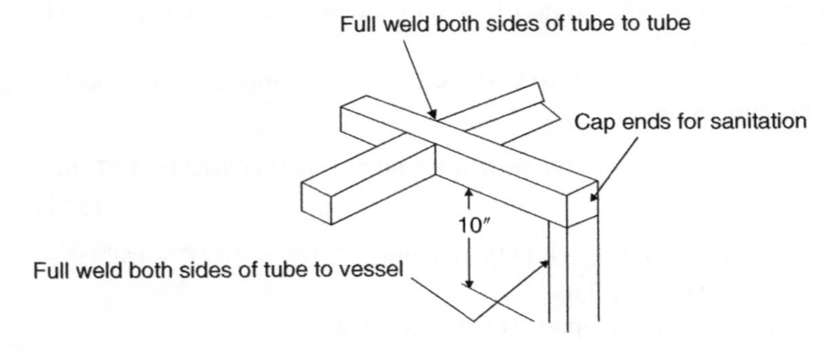

Figure 3.8 Tube reinforcing edges.

Ports: (Nozzles)

ASME Code is only required to be used and conformed to if the internal pressure in a vessel is 15 psig or higher; however, the ASME Code procedure for analyzing ports (nozzles) for reinforcing is recommended for the P_{Red} explosion relief flowing pressures normally found up to 15 psig.

Ports (nozzles) must be checked to determine the need for reinforcing the connection to the vessel. Wall material is removed to allow an opening into the vessel, and if the nozzle wall and weld areas do not provide enough replacement material, a reinforcing ring is required. No allowance for corrosion is included in the calculations and no radiographic inspection of the weld joints is assumed; therefore, the joint efficiency is assumed to be 60%.

Illustrated on Figure 3.1, there are two possible port locations. Port-1 illustrates a port on the flat head and port-2 illustrates a port on the curved cylinder wall. For the port-1 on a flat surface, the entire panel wall thickness has been considered as required to sustain the pressure in the prior stress analysis and must be considered when calculating the replacement area required. For the port-2 located on a curved cylindrical surface, the total panel wall thickness is not usually required to sustain the pressure (especially the low explosion relief pressure, P_{Red}).

The excess wall thickness not required to sustain the pressure is available for reinforcing.

The removed wall material, either cylinder wall area or flat head (roof) wall area, must be replaced by the excess nozzle wall area and/or the excess cylinder wall area determined by the calculations for the required wall thicknesses. If the required wall thicknesses are exceeded by the actual wall thicknesses, then this is the excess wall area that replaces the removed required wall area. The weld areas are also included in the available areas for reinforcing.

If the required wall thicknesses are not exceeded by the actual wall thicknesses, then a reinforcing ring is required (Figure 3.9).

$P_{\text{Red}} =$ Pressure, 5.0 psig

$S_{\text{all}} =$ Allowable stress = 21 440 psi

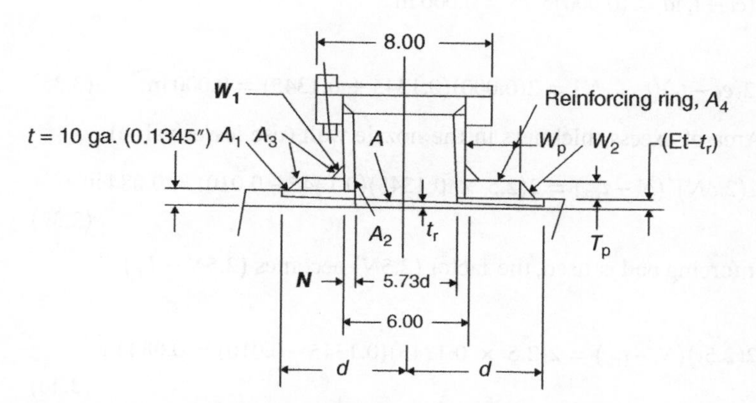

Figure 3.9 Small nozzle example 1 on a flat head. *Source:* Pressure Vessels, The ASME Code Simplified, Robert Chuse, Fifth Edition, McGraw-Hill Book Company. Figure 2.12, Pages 72 and 73. Reproduced with permission of McGraw-Hill Book Company

e = Joint efficiency = 0.60% for single weld butt joints with no backing strips (Para. UW-12 of code). Circumferential joints only, not over 5/8″ thick and not over 24″ outside diameter. No radiographic inspection of the welds is considered.

t = Actual thickness of head (in.) = 10 ga.(0.1345″) from prior assumptions

t_r = Required calculated thickness of head = t = Actual thickness of head from prior assumptions

$$Et - t_r = \text{Excess thickness in head} = 0.60 \times 0.1345 - 0.1345 = 0.000'' \tag{3.32}$$

r = Inside radius of nozzle = 5.73/2 = 2.865″

N = Actual thickness of nozzle = 0.1345″

$$t_{rn} = \text{Required calculated thickness of nozzle} = P_{\text{Red}}r/S_{\text{all}}e - 0.6P_{\text{Red}} = 5.0(2.865)/21\,440(0.6) - 0.6(5.0) = 0.0011''$$

where e = 60%joint efficiency

$$N - t_{rn} = \text{Excess thickness of nozzle} = 0.1345 - 0.0011 = 0.133'' \tag{3.33}$$

$$A_r = \text{Actual area removed} = d \times t = 5.73 \times 0.1345 = 0.771\ \text{in.}^2 \tag{3.34}$$

$$A = \text{Area of reinforcement required} = d \times t_r = 5.73 \times 0.1345 = 0.771\ \text{in.}^2 \tag{3.35}$$

$$\text{Area of reinforcement available without pad} = (A1 + A2 + A3) \tag{3.36}$$

$$A1 = \text{area of excess thickness in head (use greater value)} = (et - t_r)d = (0.000)5.73 = 0.000\ \text{in.}^2 \tag{3.37}$$

or

$$A1 = 2(et - t_r)(t + N) = 2(0.000)(0.1345 + 0.1345) = 0.000\ \text{in.}^2 \tag{3.38}$$

$A2$ = Area of excess thickness in the nozzle wall (use lesser value)

$$A2 = 2(2.5N)^*(N - t_{rn}) = 2(2.5 \times 0.1345)(0.1345 - 0.010) = 0.084\ \text{in.} \tag{3.39}$$

*If reinforcing pad is used, the factor $(2.5N)$ becomes $\left(2.5N - T_p\right)$

or

$$A2 = 2(2.5t)(N - t_{rn}) = 2(2.5 \times 0.1345)(0.1345 - 0.010) = 0.084\ \text{in.} \tag{3.40}$$

$A3$ = Cross sectional area of welds available for reinforcement. Use t_w = 1/4 in welds minimum where possible.

$$= 2\left[\left(W_1{}^2 + W_2{}^2\right)/2\right] = 2\left[\left(0.25^2 + 0.25^2\right)/2\right] = 0.125\ \text{in.}^2 \tag{3.41}$$

The area provided by $A1 + A2 + A3 = 0.000 + 0.084 + 0.125 = 0.209$ in.^{2}. (3.42)

This area is not greater than the required area of 0.771 in.^{2}; therefore, a reinforcing ring is required (Figure 3.10).

$$A4 = \text{The required area of the reinforcing ring}$$
$$A4 = 0.771 - 0.000 - 0.09 - 0.141 = 0.540\ \text{in.}^2 \tag{3.43}$$

and, $A4 = 2W_p \times T_p$where

W_p = width of reinforcing ring
T_p = thickness of reinforcing ring
$W_p = A4/T_p$

assume $T_p = 1/4''$ then

$$W_p = 0.540/0.25 = 2.16 \tag{3.44}$$

say 2-1/4″.

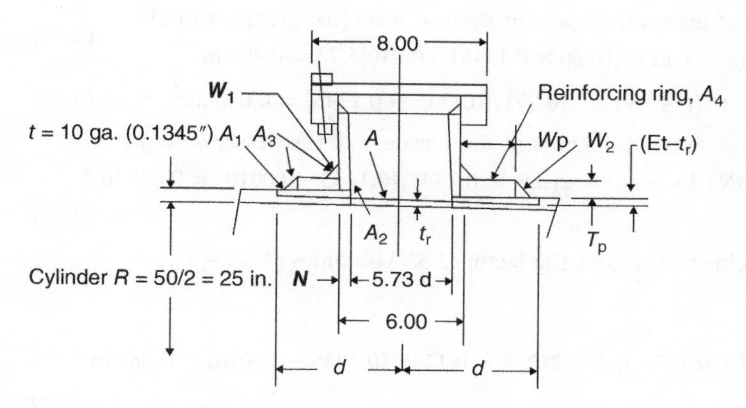

Figure 3.10 Small nozzle example 2 on a curved surface.

P_{Red} = Pressure, 5.0 psig
S_a = Allowable stress = 21 440 psi
R = Inside radius of cylinder shell = 25″
e = Joint efficiency = 0.60% for single weld butt joints with no backing strips (Para. UW-12 of code). Circumferential joints only, not over 5/8″ thick and not over 24″ outside diameter. No radiographic inspection of the welds is considered.

t_r = Required calculated thickness of shell or head

$$= P_{Red}R/S_{all}e - 0.60P_{Red} = 5.0 \times 25/21\,440 \times 0.60 - 0.60 \times 5.0 = 0.010 \text{ in.} \tag{3.45}$$

$$Et - t_r = \text{Excess thickness in shell or head} = 0.60 \times 0.1345 - 0.010 = 0.071 \text{ in.} \tag{3.46}$$

t = Actual thickness of head (in.) = 10 ga.(0.1345″)

N = Actual thickness of nozzle = 0.1345″

t_{rn} = Required calculated thickness of nozzle = $P_{Red}r/S_{all}e - 0.6P_{Red}$

$= 5.0(2.865)/21\,440(0.6) - 0.6(5.0) = 0.0011''$

where e = 60%Joint efficiency

$$N - t_{rn} = \text{Excess thickness of nozzle} = 0.1345 - 0.0011 = 0.133'' \tag{3.47}$$

$$A_r = \text{Actual area removed} = d \times t = 5.73 \times 0.1345 = 0.771 \text{ in.}^2 \tag{3.48}$$

$$A = \text{Area of reinforcement required} = d \times t_r = 5.73 \times 0.1345 = 0.771 \text{ in.}^2 \tag{3.49}$$

$$\text{Area of reinforcement available without pad} = (A1 + A2 + A3) \tag{3.50}$$

$A1$ = area of excess thickness in shell or head (use greater value)

$$= (et - t_r)d = [(0.60 \times 0.1345) - 0.010]5.73 = 0.405 \text{ in.}^2 \tag{3.51}$$

$$A1 = 2(et - t_r)(t + N) = 2(0.071)(0.1345 + 0.1345) = 0.038 \text{ in.}^2 \tag{3.52}$$

$A2$ = area of excess thickness in the nozzle wall (use lesser value)

$$= 2(2.5N)^*(N - t_{rn}) = 2(2.5 \times 0.1345)(0.1345 - 0.0010) = 0.089 \text{ in.}^2 \tag{3.53}$$

*If reinforcing pad is used, the factor $(2.5N)$ becomes $(2.5N - T_p)$

or

$$A2 = 2(2.5t)(N - t_{rn}) = 2(2.5 \times 0.1345)(0.1345 - 0.0010) = 0.089 \text{ in.}^2 \tag{3.54}$$

$A3$ = Cross sectional area of welds available for reinforcement.Use t_w

= ¼ inwelds minimum where possible.

$$= 2[(W_1^2 + W_2^2)/2] = 2[(0.25^2 + 0.25^2)/2] = 0.125 \text{ in.}^2 \tag{3.55}$$

$$\text{The area provided by } A1 + A2 + A3 = 0.405 + 0.089 + 0.125 = 0.619 \text{ in.}^2. \tag{3.56}$$

This area is not greater than the required area of 0.771 in.^{2}; therefore, a reinforcing ring is required.

$$A4 = \text{The required area of the reinforcing ring } A4 = 0.771 - 0.619 = 0.152 \text{ in.}^2 \quad (3.57)$$

and, $A4 = 2W_p \times T_p$ where

W_p = width of reinforcing ring
T_p = thickness of reinforcing ring
$W_p = A4/T_p$

assume $T_p = 1/4''$ then

$$W_p = 0.152/0.25 = 0.608 \quad (3.58)$$

say 3/4″.

Vessel Head Not Flat

Where a port (nozzle) is located on a head that is not flat, substitute one of the following equations for the required thickness, t_r, to determine the excess thickness available for reinforcing. There are two other types of heads on cylindrical vessels; an ellipsoidal head or a torispherical head.

Ellipsoidal head: refer ASME code para. UG-32(d): "The required thickness of a dished head of semi-ellipsoidal form, in which half the minor axis (inside depth of the head minus the skirt) equals one-fourth of the inside diameter of the head skirt, shall be determined by" (Figure 3.11):

$$t_r = P_{Red}D/2S_{all}e - 0.2P_{Red} \quad (3.59)$$

$$t_r = 5.0 \times 50/2 \times 21\,440 \times 0.6 - 0.2 \times 5.0 = 0.010 \text{ in.} \quad (3.60)$$

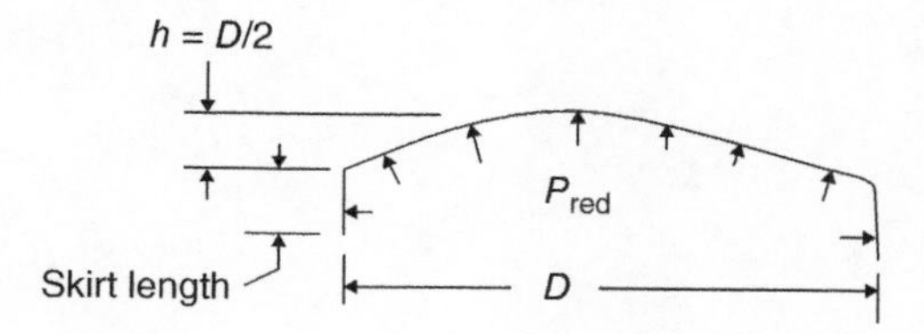

Figure 3.11 Ellipsoidal head.

Torispherical head: refer ASME code para. UG-32(e): "The required thickness of a torispherical head, in which the knuckle radius is 6% of the inside crown radius, L, shall be determined by" (Figure 3.12):

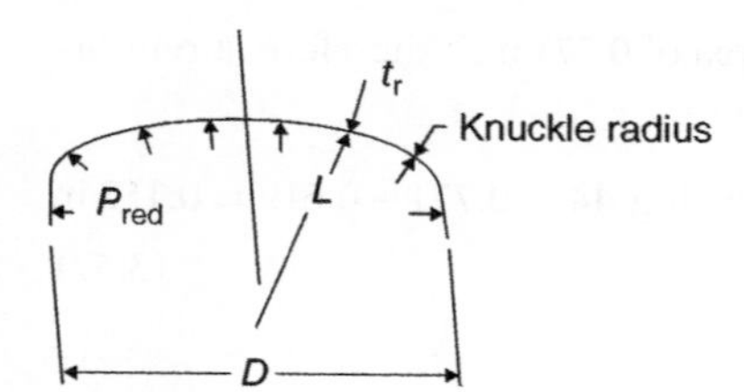

Figure 3.12 Torispherical head.

$$t_r = 0.885P_{Red}L/2S_{all}e - 0.1P_{Red} \tag{3.61}$$

where L = Inside spherical or crown radius, in.

$$t_r = 0.885 \times 5.0 \times 50/2 \times 21\,440 \times 0.6 - 0.1 \times 5.0 = 0.008\text{ in.} \tag{3.62}$$

Loads to be carried by welds and strength of connection elements:

Refer to Chapter 4 for reinforcing member to panel weld analyses and for port (nozzle) weld analyses.

4

Reinforcing Member to Panel Weld Analyses and Port (Nozzle) Weld Analyses

Warning: Clean all equipment surfaces free of dust prior to any welding.

Continuous welds are not required to safely connect the reinforcing members to the panels. Intermittent welding is used to reduce material and labor costs. Caulking of the joints between intermittent weld areas, for sanitary reasons, is recommended.

The effective length of any segment of intermittent fillet welding, according to specifications of the American Institute of Steel Construction (AISC), shall not be less than four times the weld size with a minimum of 1.5 in.

For shielded arc welds the working shear stress = 13 600 psi.

The weld area,

$$A_w = L\left(t_w/\sqrt{2}\right) = 1.5(t_w/1.414) = 1.061(t_w)\ \text{in.}^2 \tag{4.1}$$

Both legs of the angle, or both sides of the tube, channel, or rib, are welded to the panel, therefore, the two welds provide total weld area per intermittent weld.

$$2 \times A_w = 2 \times 1.061(t_w) = 2.121(t_w)\ \text{in.}^2 \tag{4.2}$$

Then the allowable load, F_a, per pair of fillet welds is (Figure 4.1)

$$13\,600\ \text{lbs/in.}^2 \times 2.121(t_w) = 28\,845(t_w)\ \text{lbs} \tag{4.3}$$

F_a = Allowable load per pair of fillet welds, lbs (refer Table 4.1)
C = Distance from centroidal axis to panel center, $Y' - 1/2t_{P}$, in.
I_C = Moment of inertia for composite section, in.^{4}
V_m = Maximum vertical shear end reaction load, total load/2
Q = Statical moment of panel about centroidal axis = A_pC
$A_p = W_pt_p$ = in.^{2}
V_h = Horizontal shear between panel and reinforcing member
= V_mQ/I_c = lb/in.
Weld pitch (maximum) = F_a/V_h = inches

Explosion Vented Equipment System Protection Guide, First Edition. Robert C. Comer.

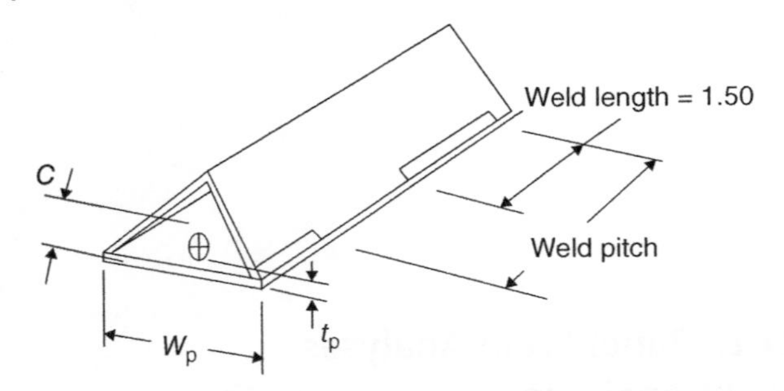

Figure 4.1 Weld spacing. F_a, allowable load per pair of fillet welds (lbs, Table 4.1); C, distance from centroidal axis to panel center = $Y' - 1/2t_p$ (in.); I_C, moment of inertia for composite section (in.^{4}); V_m, maximum vertical shear end reaction load (total load/2); Q, statistical moment of panel about centroidal axis = A_pC, $A_p = W_pt_p$ (in.^{2}); V_h, horizontal shear between panel and reinforcing member = V_mQ/I_C (lbs/in.); weld pitch (maximum) = F_a/V_h (in.).

Table 4.1 Allowable load on welds: intermittent welds occur in pairs.

Weld size	t_w (in.)	F_a (lbs)
(2) 1/8″ × 1.50″	0.125	3606
(2) 3/16″ × 1.50″	0.188	5423
(2) 1/4″ × 1.50″	0.250	7211
(2) 3/8″ × 1.50″	0.375	10 817

Refer to Tables 4.2–4.5 for values of I_C, Y', and A_p for the angle, tube, channel, or rib reinforcing.

Example 4.1

For the short reinforcing cross members: $V_m = 515.2$ lbs, from Table 4.2, 2 × 2 × 1/8 angle with 10 ga. (0.1345) thick panel:

$$I_C = 0.387 \text{ in.}^4, Y' = 0.502 \text{ in.}, A_p = 0.404 \text{ in.}^2$$

Then

$$C = 0.502 - 1/2(0.1345) = 0.433 \text{ in.} \quad (4.4)$$

$$Q = 0.404(0.433) = 0.175 \quad (4.5)$$

$$V_h = (515.2)(0.175)/0.387 = 233 \text{ lbs/in.} \quad (4.6)$$

Weld pitch = F_a/233 = for (2) 1/8 × 1.5 fillet welds = 3606/233 = 15 in. maximum. With the length of the reinforcing member only 15″, it is only required to weld 1.5″ long at the ends of the reinforcing members leaving a 12″ space between welds.

Table 4.2 Structural angle – panel composite sections: I_C, Y', A_p, and W_p.

Member angles	Parameters	10 ga. (0.1345)	8 ga. (0.1644)	7 ga. (0.1793)	3/16 (0.188)	1/4 (0.250)
2×2×1/8	I_C	0.387	0.407	0.418	0.424	0.464
	Y'	0.502	0.484	0.477	0.475	0.459
	A_p	0.404	0.493	0.538	0.564	0.75
	W_p	3.0	3.0	3.0	3.0	3.0
2×2×3/16	I_C	0.523	0.564	0.570	0.602	0.636
	Y'	0.581	0.563	0.558	0.554	0.541
	A_p	0.410	0.509	0.554	0.581	0.774
	W_p	3.094	3.094	3.094	3.094	3.094
2×2×1/4	I_C	0.648	0.686	0.703	0.715	0.720
	Y'	0.631	0.619	0.614	0.614	0.600
	A_p	0.428	0.523	0.570	0.598	0.795
	W_p	3.182	3.182	3.182	3.182	3.182
2-1/2× 2-1/2×1/4	I_C	1.271	1.336	1.367	1.384	1.509
	Y'	0.761	0.740	0.732	0.728	0.707
	A_p	0.523	0.640	0.697	0.731	0.972
	W_p	3.89	3.89	3.89	3.89	3.89
3×3×1/4	I_C	2.214	2.314	2.362	2.438	2.586
	Y'	0.896	0.862	0.852	0.845	0.811
	A_p	0.617	0.755	0.823	0.863	1.148
	W_p	4.59	4.59	4.59	4.59	4.59
3×3×3/8	I_C	2.985	3.121	3.190	3.223	3.485
	Y'	0.986	0.967	0.950	0.954	0.928
	A_p	0.642	0.784	0.856	0.897	1.193
	W_p	4.77	4.77	4.77	4.77	4.77
4×4×1/4	I_C	5.286	5.512	5.624	5.692	6.091
	Y'	1.149	1.106	1.087	1.023	1.021
	A_p	0.807	0.986	1.08	1.128	1.5
	W_p	6.0	6.0	6.0	6.0	6.0

For larger angle values not given in the table, use the "Structural Angle – Panel Worksheet" in the appendix. These values are used to determine weld pitch spacing.

Table 4.3 Structural tubing – panel composite sections: I_C, Y', A_p, and W_p.

Member tubing	Parameters	10 ga. (0.1345)	8 ga. (0.1644)	7 ga. (0.1793)	3/16 (0.188)	1/4 (0.250)
2×2×3/16	I_C	0.907	1.032	1.062	1.083	1.201
	Y'	0.910	0.900	0.894	0.892	0.879
	A_p	0.336	0.410	0.448	0.470	0.625
	W_p	2.50	2.50	2.50	2.50	2.50
2×2×1/4	I_C	1.124	1.209	1.264	1.282	1.337
	Y'	0.948	0.942	0.939	0.938	0.932
	A_p	0.336	0.411	0.448	0.470	0.625
	W_p	2.50	2.50	2.50	2.50	2.50
2-1/2×2-1/2 ×1/4	I_C	2.338	2.489	2.565	2.61	2.938
	Y'	1.171	1.154	1.154	1.152	1.137
	A_p	0.404	0.493	0.538	0.564	0.75
	W_p	3.00	3.00	3.00	3.00	3.00
3×3×1/4	I_C	4.139	4.338	4.438	4.495	4.892
	Y'	1.393	1.377	1.369	1.365	1.339
	A_p	0.471	0.575	0.628	0.658	0.875
	W_p	3.50	3.50	3.50	3.50	3.50
4×3×3/16	I_C	6.912	7.237	7.402	7.494	7.861
	Y'	1.794	1.761	1.745	1.736	1.680
	A_p	0.471	0.574	0.628	0.658	0.875
	W_p	3.50	3.50	3.50	3.50	3.50
4×3×1/4	I_C	8.199	8.552	8.732	8.830	9.834
	Y'	1.861	1.837	1.826	1.820	1.781
	A_p	0.471	0.575	0.628	0.658	0.875
	W_p	3.50	3.50	3.50	3.50	3.50
4×3×5/16	I_C	9.356	9.610	9.798	9.905	10.656
	Y'	1.903	1.886	1.878	1.874	1.846
	A_p	0.471	0.575	0.628	0.658	0.875
	W_p	3.50	3.50	3.50	3.50	3.50
4×4×1/4	I_C	10.433	10.880	11.098	11.222	11.292
	Y'	1.836	1.808	1.796	1.789	1.743
	A_p	0.605	0.740	0.807	0.846	1.125
	W_p	4.50	4.50	4.50	4.50	4.50

Table 4.3 (Continued)

Member tubing	Parameters	10 ga. (0.1345)	8 ga. (0.1644)	7 ga. (0.1793)	3/16 (0.188)	1/4 (0.250)
4×4×3/8	I_C	13.010	13.501	13.741	13.882	14.865
	Y'	1.914	1.900	1.893	1.889	1.862
	A_p	0.605	0.740	0.807	0.846	1.125
	W_p	4.50	4.50	4.50	4.50	4.50
5×3×1/4	I_C					16.151
	Y'					2.235
	A_p					0.875
	W_p					3.50

For larger tube values not given in the table, use the "Structural Tube – Panel Worksheet" in the appendix. These values are used to determine weld pitch spacing.

Table 4.4 Structural channel – panel composite sections: I_C, Y', A_p, and W_p.

Member channel	Parameters	10 ga. (0.1345)	8 ga. (0.1644)	7 ga. (0.1793)	3/16 (0.188)	1/4 (0.250)
C3×4.1	I_C	0.399	0.440	0.459	0.471	0.551
	Y'	0.623	0.615	0.612	0.608	0.606
	A_p	0.471	0.575	0.628	0.658	0.875
	W_p	3.50	3.50	3.50	3.50	3.50
C3×5	I_C	0.487	0.533	0.522	0.571	0.672
	Y'	0.685	0.679	0.642	0.676	0.673
	A_p	0.471	0.575	0.628	0.658	0.875
	W_p	3.50	3.50	3.50	3.50	3.50
C3×6	I_C	0.585	0.642	0.672	0.688	0.808
	Y'	0.749	0.745	0.743	0.743	0.741
	A_p	0.471	0.575	0.628	0.658	0.875
	W_p	3.00	3.00	3.00	3.00	3.00
C4×5.4	I_C	0.643	0.707	0.715	0.755	0.875
	Y'	0.689	0.678	0.675	0.672	0.662
	A_p	0.605	0.740	0.807	0.846	1.125
	W_p	4.50	4.50	4.50	4.50	4.50
C4×7.25	I_C	0.839	0.923	0.962	0.986	1.153
	Y'	0.789	0.782	0.778	0.776	0.770

(Continued)

Table 4.4 (Continued)

Member channel	Parameters	10 ga. (0.1345)	8 ga. (0.1644)	7 ga. (0.1793)	3/16 (0.188)	1/4 (0.250)
	A_p	0.605	0.740	0.807	0.846	1.125
	W_p	4.50	4.50	4.50	4.50	4.50
C5 × 6.7	I_C	0.956	1.048	1.081	1.119	1.295
	Y'	0.752	0.738	0.726	0.729	0.714
	A_p	0.740	0.904	0.986	1.034	1.375
	W_p	5.50	5.50	5.50	5.50	5.50
C5 × 9	I_C	1.223	1.341	1.399	1.432	1.663
	Y'	0.856	0.845	0.841	0.838	0.826
	A_p	0.740	0.904	0.986	1.034	1.375
	W_p	5.50	5.50	5.50	5.50	5.50

For larger channel values not given in the table, use the "Structural Channel – Panel Worksheet" in the appendix. These values are used to determine weld pitch spacing.

Table 4.5 Rib – panel composite sections: I_C, Y', A_p, and W_p.

Member rib	Parameters	10 ga. (0.1345)	8 ga. (0.1644)	7 ga. (0.1793)	3/16 (0.188)	1/4 (0.250)
1/4 × 2.0	I_C	0.275	0.295	0.308	0.318	0.361
	Y'	0.957	0.950	0.949	0.947	0.942
	A_p	0.101	0.123	0.134	0.141	0.188
	W_p	0.75	0.75	0.75	0.75	0.75
1/4 × 3.0	I_C	0.535	0.575	0.591	0.602	0.673
	Y'	1.449	1.441	1.437	1.435	1.424
	A_p	0.101	0.123	0.134	0.141	0.188
	W_p	0.75	0.75	0.75	0.75	0.75
1/4 × 4.0	I_C	0.887	0.933	0.957	1.017	1.070
	Y'	1.914	1.936	1.932	1.901	1.914
	A_p	0.101	0.123	0.134	0.141	0.188
	W_p	0.75	0.75	0.75	0.75	0.75
1/4 × 5.0	I_C	1.311	1.377	1.406	1.425	1.549
	Y'	2.459	2.433	2.428	2.425	2.408
	A_p	0.101	0.123	0.134	0.141	0.188
	W_p	0.75	0.75	0.75	0.75	0.75
1/4 × 6.0	I_C	1.836	1.964	1.939	1.962	2.113
	Y'	2.987	2.931	2.926	2.922	2.902
	A_p	0.101	0.123	0.134	0.141	0.188
	W_p	0.75	0.75	0.75	0.75	0.75

Table 4.5 (Continued)

Member rib	Parameters	10 ga. (0.1345)	8 ga. (0.1644)	7 ga. (0.1793)	3/16 (0.188)	1/4 (0.250)
3/8 × 2.0	I_C	0.354	0.396	0.408	0.415	0.468
	Y'	1.197	0.990	0.990	0.990	0.996
	A_p	0.118	0.144	0.157	0.164	0.218
	W_p	0.875	0.875	0.875	0.875	0.875
3/8 × 3.0	I_C	0.745	0.784	0.806	0.813	0.808
	Y'	1.486	1.484	1.479	1.485	1.485
	A_p	0.118	0.144	0.157	0.164	0.218
	W_p	0.875	0.875	0.875	0.875	0.875
3/8 × 4.0	I_C	1.244	1.297	1.323	1.339	1.447
	Y'	1.984	1.982	1.981	1.980	1.979
	A_p	0.118	0.144	0.157	0.164	0.218
	W_p	0.875	0.875	0.875	0.875	0.875
3/8 × 5.0	I_C	1.868	1.937	1.967	1.985	2.125
	Y'	2.482	2.480	2.479	2.479	2.475
	A_p	0.118	0.144	0.157	0.164	0.218
	W_p	0.875	0.875	0.875	0.875	0.875
3/8 × 6.0	I_C	2.617	2.697	2.737	2.758	2.926
	Y'	2.982	2.979	2.978	2.978	2.973
	A_p	0.118	0.144	0.157	0.164	0.218
	W_p	0.875	0.875	0.875	0.875	0.875
1/2 × 2.0	I_C	0.464	0.492	0.507	0.515	0.573
	Y'	1.008	1.012	1.013	1.015	1.025
	A_p	0.134	0.164	0.179	0.188	0.25
	W_p	1.00	1.00	1.00	1.00	1.00
1/2 × 3.0	I_C	0.949	0.992	1.012	1.027	1.114
	Y'	1.506	1.508	1.510	1.512	1.517
	A_p	0.134	0.164	0.179	0.188	0.25
	W_p	1.00	1.00	1.00	1.00	1.00
1/2 × 4.0	I_C	1.600	1.658	1.686	1.703	1.822
	Y'	2.004	2.006	2.008	2.008	2.014
	A_p	0.134	0.164	0.179	0.188	0.25
	W_p	1.00	1.00	1.00	1.00	1.00

For larger rib values not given in the table, use the "Rib – Panel Worksheet" in the appendix. These values are used to determine weld pitch spacing.

Example 4.2

For the short reinforcing cross members: $V_{\mathrm{m}} = 802$ lbs, from Table 4.2, 2 × 2 × 1/8 angle with 10 ga. (0.1345) thick panel:

$$I_{\mathrm{C}} = 0.387\ \text{in.}^4,\quad Y' = 0.502\ \text{in.}, A_{\mathrm{p}} = 0.404\ \text{in.}^2$$

Then

$$C = 0.502 - 0.067 = 0.433\ \text{in.} \tag{4.7}$$

$$Q = 0.175$$

$$V_{\mathrm{h}} = 802(0.175)/0.387 = 363\ \text{lbs/in.} \tag{4.8}$$

Weld pitch = for (2) 1/8 × 1.5 fillet welds = 3606/363 = 9.9 in. maximum. With the length of the short reinforcing members 19.55″, the ends are to be welded (2) 1/8″ × 1.5″ long and in the center add (2) 1/8″ × 1.5″ long welds.

From Example 4.2

For the main reinforcing member: $V_{\mathrm{m}} = 3208$ lbs, from Table 4.2, 3 × 3 × 3/8 angle with 10 ga. (0.1345) thick panel:

$$I_{\mathrm{C}} = 2.985\ \text{in.}^4, Y' = 0.986\ \text{in.}, A_{\mathrm{p}} = 0.642\ \text{in.}^2$$

Then

$$C = 0.986 - 0.067 = 0.919\ \text{in.} \tag{4.9}$$

$$Q = 0.642(0.919) = 0.590 \tag{4.10}$$

$$V_{\mathrm{h}} = 3208(0.590)/2.985 = 634\ \text{lbs/in.} \tag{4.11}$$

$$\text{Weld pitch} = \text{for (2) } 1/8 \times 1.5 \text{ fillet welds} = 3606/634 = 5.68\ \text{in. maximum} \tag{4.12}$$

It is obvious that the 1/8″ fillet welds are not strong enough to sustain the load. Try (2) 1/4″ × 1.5″ fillet welds. From Table 4.1, $F_{\mathrm{a}} = 7211$ lbs.

$$\text{Weld pitch} = 7211/634 = 11.4\ \text{in.maximum} \tag{4.13}$$

Use a weld pitch of 11.0 in. This will provide nine equally spaced (2) 1/4″ × 1.5″ long welds along the 97.75″ length of the member.

Corner Miter Joint Reinforcing Weld Analysis

Refer Figure 3.3 for 2″ × 1/4″ rib (Figure 4.2).

$$L = \text{weld length (in.)} = (2)2''/\sin 45^\circ = 2 \times 2/0.7071 = 5.657\ \text{in.} \tag{4.14}$$

$$\begin{aligned} R &= 1109\ \text{in.} \\ R_{\mathrm{n}} &= 1109/\sin 45^\circ = 1109/0.7071 = 1568\ \text{lbs} \end{aligned} \tag{4.15}$$

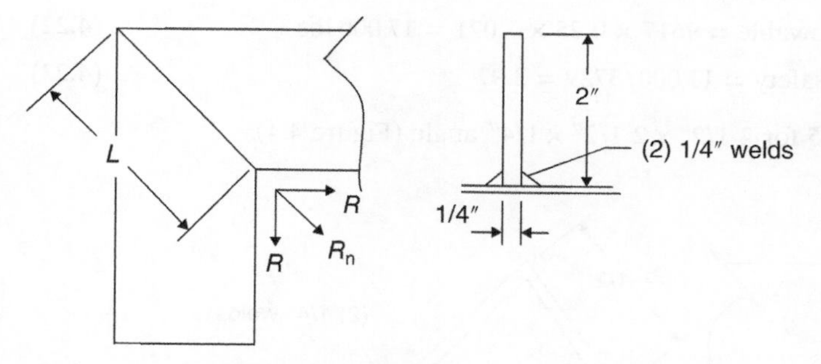

Figure 4.2 Head reinforcing rib member 1.

$$\text{Allowable shear in weld} = 13\,600(L \times t \times \sin\ 45°) = 13\,600(L \times t \times 0.7071) = 9617 \times t/\text{inch of weld} \quad (4.16)$$

$$\text{Then } R \text{ allowable} = 9617 \times 0.25 \times 5.657 = 13\,600 \text{ lbs} \quad (4.17)$$

$$\text{Factor of safety} = 13\,600/1568 = 8.67 \quad (4.18)$$

Refer Figure 3.5 for 2 1/2″ × 2 1/2″ × 1/4″ tube (Figure 4.3).

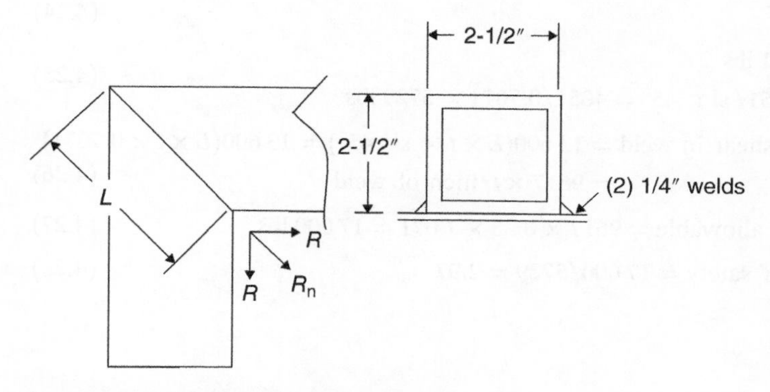

Figure 4.3 Head reinforcing member 3.

$$L = \text{weld length (in.)} = (2)\ 2 - 1/2/\sin 45° = 2 \times 2 - 1/2/0.7071 = 7.071 \text{ in.} \quad (4.19)$$

$$R = 4051 \text{ lbs}$$

$$R_n = 4051/\sin\ 45° = 4051/0.7071 = 5729 \text{ lbs} \quad (4.20)$$

$$\text{Allowable shear in weld} = 13\,600(L \times t \times \sin\ 45°) = 13\,600(L \times t \times 0.7071) = 9617 \times t/\text{inch of weld} \quad (4.21)$$

$$\text{Then } R \text{ allowable} = 9617 \times 0.25 \times 7.071 = 17\,000 \text{ lbs} \tag{4.22}$$

$$\text{Factor of safety} = 17\,000/5729 = 2.97 \tag{4.23}$$

Refer Figure 3.5 for 2 1/2″ × 2 1/2″ × 1/4″ angle (Figure 4.4).

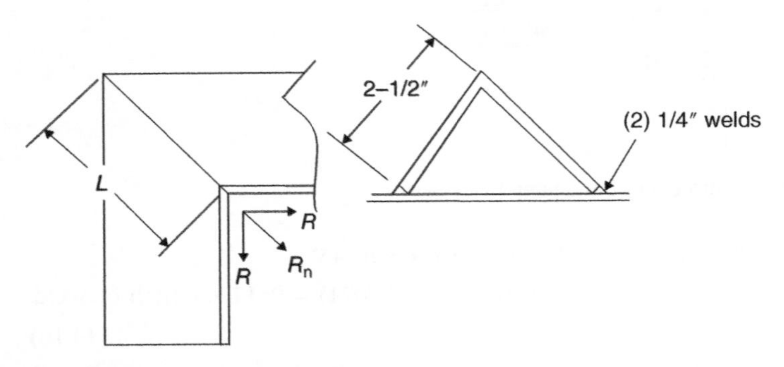

Figure 4.4 Head reinforcing angle member 3.

$$L = \text{Weld length (in.)} = 2(2 - 1/2/\sin 45°) = 2 \times 2 - 1/2/0.7071 = 7.071 \text{ in.} \tag{4.24}$$

$$R = 4051 \text{ lbs}$$
$$R_n = 4051/\sin\ 45° = 4051/0.7071 = 5729 \text{ lbs} \tag{4.25}$$

$$\text{Allowable shear in weld} = 13\,600(L \times t \times \sin 45°) = 13\,600(L \times t \times 0.7071)$$
$$= 9617 \times t/\text{inch of weld} \tag{4.26}$$

$$\text{Then } R \text{ allowable} = 9617 \times 0.25 \times 7.071 = 17\,000 \text{ lbs} \tag{4.27}$$

$$\text{Factor of safety} = 17\,000/5729 = 2.97 \tag{4.28}$$

Port (Nozzle) Reinforcing Weld Analyses

Refer to Figure 3.9: Small nozzle example 1 on a flat head.

Load to be carried by welds: (para. UG-41(b)(1)). Refer to ASME Boiler and Pressure Vessel Code, Unfired Pressure Vessels, Section V111, in accordance with Division 1 design basis.

$$W = (A - A_1)S_a = (0.771 - 0.000)21\,440 = 16\,530 \text{ lbs required strength} \tag{4.29}$$

Unit weld stresses: (para. UW-15(b)). The allowable stresses for groove and fillet welds and nozzle neck shear as a percentage of vessel material:

$$\text{Fillet weld shear} = 0.49 \times 21\,440 = 10\,505 \text{ psi} \tag{4.30}$$

$$\text{Groove weld tension} = 0.74 \times 21\,440 = 15\,866 \text{ psi} \tag{4.31}$$

$$\text{Groove weld shear} = 0.60 \times 21\,440 = 12\,864 \text{ psi} \tag{4.32}$$

$$\text{Nozzle wall shear} = 0.70 \times 21\,440 = 15\,008 \text{ psi} \tag{4.33}$$

Strength of connected elements

A) $\text{Fillet weld shear} = \pi/2 \times \text{nozzle O.D.} \times \text{weld leg} \times 10\,505 = 1.57 \times 6.00 \times 0.25 \times 10\,505 = 24\,739 \text{ lbs}$ (4.34)

B) $\text{Nozzle wall shear} = \pi/2 \times \text{mean nozzle diameter} \times t_n \times 15008 = 1.57 \times (5.73 + 6.00)/2 \times 0.1345 \times 15\,008 = 18\,587 \text{ lbs}$ (4.35)

C) $\text{Groove weld tension} = \pi/2 \times \text{nozzle O.D.} \times \text{weld leg} \times 15\,866 = 1.57 \times 6.00 \times 0.25 \times 15\,866 = 37\,364 \text{ lbs}$ (4.36)

D) $\text{Outer fillet weld shear} = \pi/2 \times \text{reinforcing ring O.D.} \times \text{weld leg} \times 10\,505 = 1.57 \times (6.00 + 4.50) \times 0.25 \times 10\,505 = 43\,293 \text{ lbs}$ (4.37)

Possible paths of failure

1) Through (B) and (D) $= 18\,587 + 43\,293 = 61\,880$ lbs (4.38)

2) Through (A) and (C) $= 24\,793 + 37\,364 = 62\,157$ lbs (4.39)

3) Through (C) and (D) $= 37\,364 + 43\,293 = 80\,657$ lbs (4.40)

All are stronger than the required strength of 16 530 lbs.
Refer to Figure 3.10: Small nozzle example 2 on a curved surface.
Load to be carried by welds: (para. UG-41(b)(1))

$$W = (A - A_1)S_a = (0.771 - 0.124)21\,440 = 13\,871 \text{ lbs required strength}$$

Unit weld stresses: (para. UW-15(b)). The allowable stresses for groove and fillet welds and nozzle neck shear as a percentage of vessel material:

$$\text{Fillet weld shear} = 0.49 \times 21\,440 = 10\,505 \text{ psi} \tag{4.41}$$

$$\text{Groove weld tension} = 0.74 \times 21\,440 = 15\,866 \text{ psi} \tag{4.42}$$

$$\text{Groove weld shear} = 0.60 \times 21\,440 = 12\,864 \text{ psi} \tag{4.43}$$

$$\text{Nozzle wall shear} = 0.70 \times 21\,440 = 15\,008 \text{ psi} \tag{4.44}$$

Strength of connected elements

A) Fillet weld shear = $\pi/2 \times$ nozzle O.D. $\times$ weld leg $\times$ 10 505

$$= 1.57 \times 6.00 \times 0.25 \times 10\,505 = 24\,739 \text{ lbs} \tag{4.45}$$

B) Nozzle wall shear = $\pi/2 \times$ mean nozzle diameter $\times\, t_n \times$ 15 008

$$= 1.57(5.73 + 6.00) \times 0.1345 \times 15\,008 = 18\,587 \text{ lbs} \tag{4.46}$$

C) Groove weld tension = $\pi/2 \times$ nozzle O.D. $\times$ weld leg $\times$ 15 866

$$= 1.57 \times 6.00 \times 0.25 \times 15\,866 = 37\,364 \text{ lbs} \tag{4.47}$$

D) Outer fillet weld shear = $\pi/2 \times$ reinforcing ring O.D. $\times$ weld leg $\times$ 10 505

$$= 1.57(6.00 + 4.50)0.25 \times 10\,505 = 43\,293 \text{ lbs} \tag{4.48}$$

Possible paths of failure

1) Through (B) and (D) = 18 587 + 43 293 = 61 880 lbs (4.49)

2) Through (A) and (C) = 24 793 + 37 364 = 62 157 lbs (4.50)

3) Through (C) and (D) = 37 364 + 43 293 = 80 657 lbs (4.51)

All are stronger than the required strength of 13 871 lbs.

5

Access Doors, Hinges, and Latches

Access Door

Figure 2.1 illustrates a typical access door configuration. The door is 37-7/8″ × 40″ × 10 ga. (0.1345″). The door is supported by three hinges and three latches. Structural tubing is recommended for the reinforcing members on access doors to provide flat surfaces for latches and hinges to be attached.

The analysis as described in Chapter 2 for panel stresses with $P_{Red} = 5$ psig and a panel thickness of 10 ga. also applies here, and the maximum panel size is 19″ × 19″ square. By dividing the 37-7/8″ width in half yields a panel width of 18.94″. By dividing the 40″ height in half yields a panel height of 20″. As described in Chapter 2, when the reinforcing members are added, the actual unsupported panel size is reduced by the width of the reinforcing member; therefore, slightly exceeding the 19″ × 19″ spacing of the reinforcing members is acceptable.

The load per inch "W" on the reinforcing member is the reaction of two adjacent panels; therefore, the uniform load is assumed to be twice the maximum reaction (Figure 5.1).

$$R_{max} = \gamma P_{Red} b \text{ for each panel. From Table 2.1, for } a/b = 20/18.94 = 1.0,\ \Upsilon = 0.420 \tag{5.1}$$

$$\text{Then, } R_{max} = 0.420(5.0)18.94 = 39.77\ \text{lbs/in. (per panel)} \tag{5.2}$$

$$W = 2R_{max} = 2(39.77) = 79.54\ \text{lbs/in.} \tag{5.3}$$

$$M_{max} = \text{Max moment} = WL^2/8 = 79.54\left(18.94^2\right)/8 = 3566\ \text{lbs} \tag{5.4}$$

$$R_l = R_r = 79.54(18.94)/2 = 753\ \text{lbs} \tag{5.5}$$

$$F_{max} = M_{max}/S_c = 32\,400\ \text{psi maximum allowable stress} \tag{5.6}$$

Explosion Vented Equipment System Protection Guide, First Edition. Robert C. Comer.

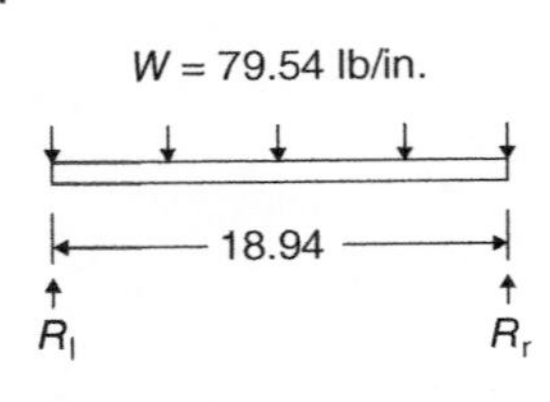

Figure 5.1 Panel reinforcing loading.

The required minimum section modulus S_c for the member is then:

$$S_c = 3566/32\,400 = 0.110\ \text{in.}^3 \tag{5.7}$$

Use a structural tube to provide a flat surface for ease of mounting hinges and latches.

From Table 2.4, the smallest recommended structural tube is 2″ × 2″ × 3/16″, and under the 10 ga. (0.1345″) panel column, the section modulus is $S_c = 0.996$ in.^{3}

The vertical door divider is 40 in. long and will be the main reinforcing member in support of the cross member angles (Figure 5.2).

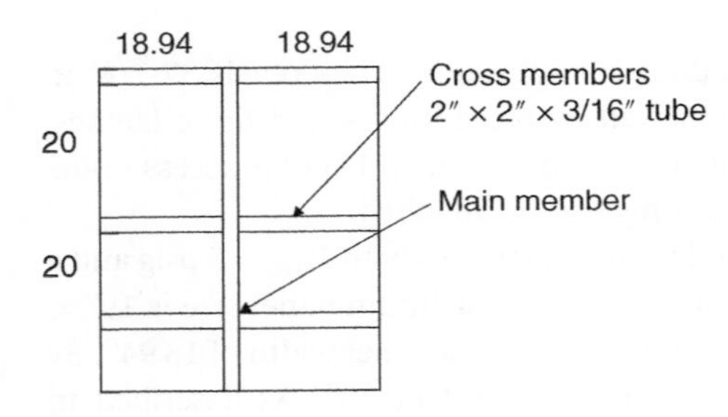

Figure 5.2 Main reinforcing member for access door.

The end reaction from each cross member Is 753 lbs. The load on the main member is 753 × 2 = 1506 lbs applied at the center of the main member. The end reactions on the main member are 1506/2 = 753 lbs. The maximum moment induced in the main member is $M_{max} = 1506 \times 20'' = 30\,120$ lbs (5.8)

The required minimum section modulus S_c is $30\,120/32\,400 = 0.930$ in.^{3} (5.9)

From Table 2.4, the smallest recommended tube under the 10 ga. (0.1345″) panel is 2″ × 2″ × 3/16″ with a section modulus of 0.996 in.^{3}

Reinforcing Member to Access Door Weld Analysis

Continuous welds are not required to safely connect the reinforcing members to the panels. The effective length of any segment of intermittent fillet welding,

Figure 5.3 Weld spacing.

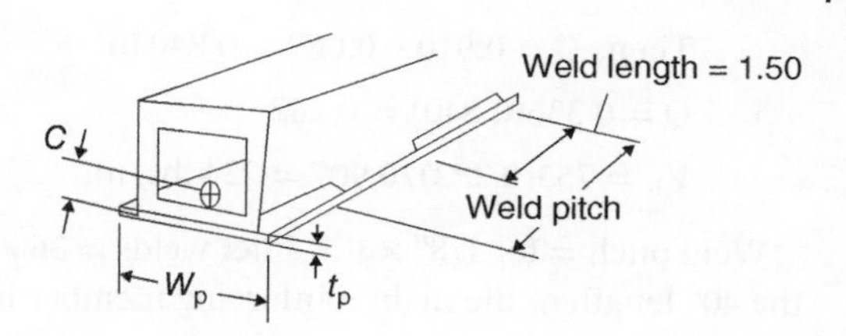

according to specifications of AISC, shall not be less than four times the weld size with a minimum of 1.5 in. (Figure 5.3).

F_a = Allowable load per pair of fillet welds, lbs (refer Table 4.1) for 1/8″ × 1.5″ = 3606 lbs.
C = Distance from centroidal axis to panel center, in.
I_C = Moment of inertia for composite section, in.^{4}
V_m = Maximum vertical shear end reaction load, total load/2.
Q = Statical moment of panel about centroidal axis = A_pC.

$$A_p = W_p t_p = \text{in.}^2$$

V_h = Horizontal shear between panel and reinforcing member $V_m Q/I_c$ = lbs/in.

$$\text{Weld pitch (maximum)} = F_{all}/V_h = \text{inches}$$

Refer to Tables 4.2, 4.3, 4.4, and 4.5 for values of I_C, Y', and A_p.

For the access door: For the short reinforcing cross members: V_m = 753 lbs. From Table 4.3 2 × 2 × 3/16 tube with 10 ga. (0.1345) thick panel:

$$I_c = 0.907\ \text{in.}^4, Y' = 0.910\ \text{in.}, A_p = 0.336\ \text{in.}^2$$

$$\text{Then, } C = 0.907 - 1/2(0.1345) = 0.840\ \text{in.}$$

$$Q = 0.336(0.840) = 0.282$$

$$V_h = 753(0.282)/0.907 = 234\ \text{lbs/in.}$$

$$\text{Weld pitch} = F_{all}/234 = \text{for } 1/8'' \times 1.5'' \text{fillet welds} = 3606/234 = 15.4\ \text{in. maximum} \tag{5.10}$$

With the length of the reinforcing cross member only 18.94″, it is required to weld the ends of the reinforcing members and one additional weld pair in the center.

For the access door main reinforcing member: V_m = 753 lbs, from Table 4.3, 2 × 2 × 3/16 tube with 10 ga. (0.1345) thick panel:

$$I_c = 0.907\ \text{in.}^4, Y' = 0.910\ \text{in.}, A_p = 0.336\ \text{in.}^2$$

Then, $C = 0.910 - 0.067 = 0.840$ in.

$Q = 0.336(0.840) = 0.282$

$$V_h = 753(0.282)/0.907 = 234 \text{ lbs/in.} \tag{5.11}$$

Weld pitch = for 1/8″ × 1.5″ fillet welds = 3606/234 = 15.41″ maximum. Divide the 40″ length of the main reinforcing member into three sections of 40/3 = 13.3″ per section.

Hinges and Latches

There are many hinge and latch designs. A couple of typical latch designs are analyzed to give guidance. Some are bolted and some are held with pull action toggle clamps (DeStaco Co.) (Tables 5.1 and 5.2).

Bolted Configuration

Refer to Figure 5.4 for a standard bolted latch configuration with three latches and three hinges for this example. The bolts are 1/2″ diameter.

Table 5.1 Bolt material and 67% yield strength.

ASTM designation	Yield strength (psi min)	67% yield strength (psi)
A 36	36 000	24 120
A 307	36 000	24 120
A 325	92 000	61 640
A 354, Gr. BD	130 000	87 100
A354, Gr. BC	109 000	73 030
A 449	92 000	61 640
A 490	130 000	87 100
A 572, Gr. 50	50 000	33 500
A 572, Gr. 42	42 000	28 140
A 588	50 000	33 500
A 687	105 000	70 350

Note: If bolt material is not specified, use 67% of 30 000 psi yield strength = 20 100 psi (lowest yield strength material).
Source: Based on AISC "Manual of Steel Construction, 8th Edition" (1980).

Table 5.2 Bolt sizes and stress areas.

Nominal bolt size	Course thread (threads per inch)	Course thread stress area (in.^{2})	Fine thread (threads per inch)	Fine thread stress area (in.^{2})
1/4	20	0.0317	28	0.0362
5/16	18	0.0522	24	0.0579
3/8	16	0.0773	24	0.0876
7/16	14	0.1060	20	0.1185
1/2	13	0.1416	20	0.1597
1/2	12	0.1374		
9/16	12	0.1816	18	0.2026
5/8	11	0.2256	18	0.2555
3/4	10	0.3340	16	0.3724
7/8	9	0.4612	14	0.5088
1.0	8	0.6051	12	0.6624
1-1/8	7	0.7627	12	0.8549
1-1/4	7	0.9684	12	1.0721
1-3/8	6	1.1538	12	1.3137
1-1/2	6	1.4041	12	1.5799

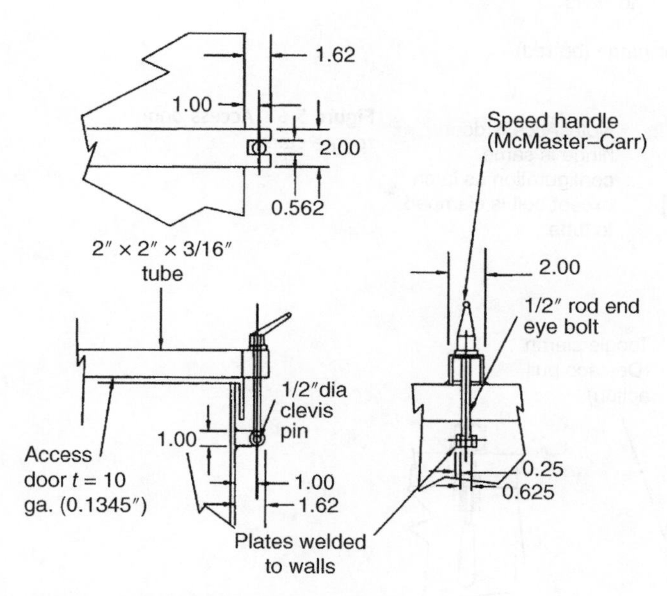

Figure 5.4 Access door latch (bolted).

$$F = \text{Total load on the access door} = P_{\text{Red}}A_{\text{door}} = 5.0(37.875 \times 40) = 7575\ \text{lbs} \tag{5.12}$$

Load per hinge and latch = 7575/6 = 1263 lbs. The preload on the eyebolts from tightening the handle is approximately 160 in.-lbs (Standard for hand tightening) (Figures 5.5 and 5.6).

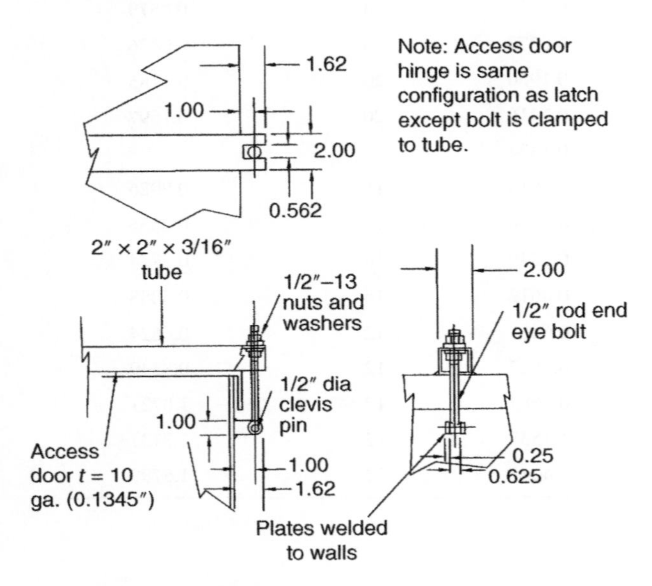

Figure 5.5 Access door hinge (bolted).

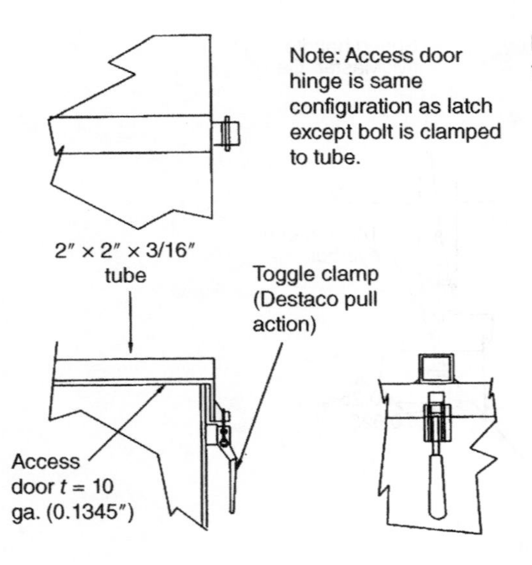

Figure 5.6 Access door toggle clamp.

The eyebolt axial load due to the tightening torque is F_T = Torque/0.2 × Nominal diameter of bolt 160/0.20 × 0.5 = 1600 lbs (Vallance and Daughtie 1951). The preload on the eyebolt due to tightening is higher than the pressure load on the access door.

The eyebolt stress is F_T/Stress area =1600/0.1416 in.^{2} = 11 299 psi (a very low stress on the eyebolt) (Figures 5.7 and 5.8).

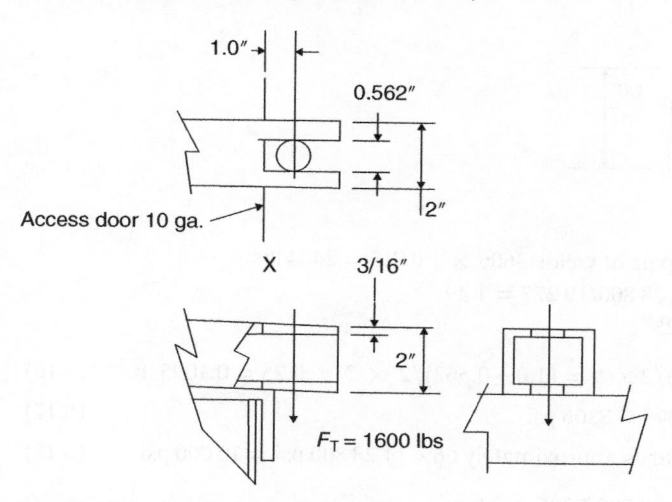

Figure 5.7 Hinge/latch reinforcing tube.

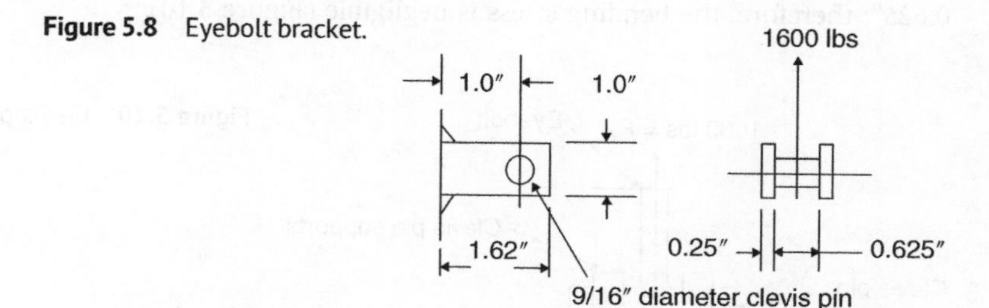

Figure 5.8 Eyebolt bracket.

$$M_{max} = F_T(1.0) = 1600 \text{ lbs}$$

$$S_c = 0.996 \text{ in.}^3 \text{ (refer Table 2.4) for } 2'' \times 2'' \times 3/16'' \text{ tube and 10 ga. door.}$$

$$f_{max} = M/S_c = 1600/0.996 = 1606 \text{ psi (very safe)} \tag{5.13}$$

$$M_{max} = F_T(1.0) = 1600(1.0) = 1600 \text{ in.-lbs}$$

$$I = bh^3/12 = 0.25 \times 2 \times 1.0^3/12 = 0.0417 \text{ in.}^4$$

$$C = 1.0/2 = 0.50, \text{ then } S = I/C = 0.0417/0.50 = 0.083 \text{ in.}^3 \tag{5.14}$$

$$f_{max} = M/S = 1600/0.083 = 19\,277 \text{ psi}$$

$$\text{Factor of safety} = 28\,800/19\,277 = 1.49 \tag{5.15}$$

Welds

For 1/8″ welds, the allowable load per pair of fillet welds is 3606 for 1.50 long welds (refer Table 4.1 and Figure 5.9).

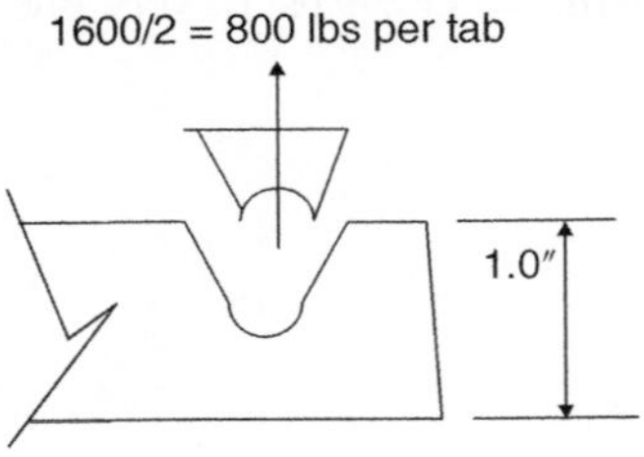

1600/2 = 800 lbs per tab

1.0″

Figure 5.9 Tear out analysis.

For the 1.0″ long pair of welds 3606 × 1.0/1.5 = 2404 lbs
Factor of safety = 28 800/19 277 = 1.49
$D = 0.562''$, $t = 0.25''$

$$A_s = (1.0 - D)/2 \times 2t = (1.0 - 0.562)/2 \times 2 \times 0.25 = 0.1095\ \text{in.}^2 \quad (5.16)$$

$$F_s = 800/0.1095 = 7306\ \text{psi} \quad (5.17)$$

Allowable shear is approximately 66% of 28 800 psi = 19 000 psi (5.18)

Factor of safety = 19 000/7306 = 2.6
The clevis pin is in double shear, and the span between the clevis pin supports is 0.625″; therefore, the bending stress is negligible (Figure 5.10).

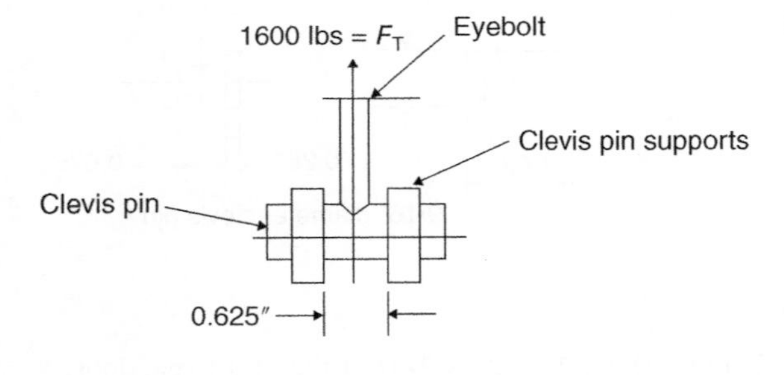

Figure 5.10 Clevis pin.

$$A_p = \text{The pin area} = \pi(D^2)/4 = 3.142(0.562^2)/4 = 0.248\ \text{in.}^2 \times 2\ \text{shear planes} \quad (5.19)$$

$$f_s = F_T/2A_p = 1600/(2 \times 0.248) = 3225\ \text{psi} \quad (5.20)$$

$$\text{Factor of safety} = 19\,000/3225 = 5.89 \quad (5.21)$$

Bending of the fastener may be the criteria, rather than shear if oversized holes are used (Figure 5.11).

Maximum bending moment, M, is:

$$M = F/2(b/3 + a/4)$$

And, $M = f_t \times I/C$ where, I = moment of inertia of bolt = $\pi d^4/64$ and $C = d/2$

By substitution: $Fb/6 + Fa/8 = \pi d^3 f_t/32$. $a = 0.625''$, $b = 0.25''$ and $f_t = 28\,800$ psi for a bolt allowable stress (no threads in the stress area). Then,

$$F\,\text{allowable} = 3\pi d^3 f_t/4(3a + 4b) = 3\pi(0.50^3) \times 28\,800/4(3 \times 0.25 + 4 \times 0.625$$
$$= 2610\,\text{lbs allowable bending load} \quad (5.22)$$

$$\text{Factor of safety} = 2610/1600 = 1.63 \quad (5.23)$$

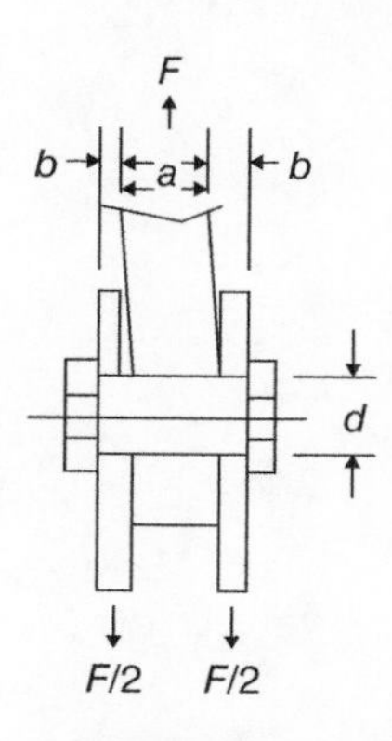

Figure 5.11 Bending analysis.

Toggle Clamp Configuration

The toggle clamps are rated for axial restraint on the access door (Figure 5.6). For this example, the toggle clamp must have a minimum rating of 1263 lbs. (The load due to pressure of 5 psig on the access door divided by six restraints) The closest DeStaco toggle clamp rating is 2000 lbs. Model no. 341. For other applications, refer to DeStaco Toggle Clamps from Dal-Tex Enterprises, Inc. Decatur, Texas.

6

Explosion Vent Ducts, Mill Inlet Air Ducts, Blast Deflectors, and Filter Bag Cage Design

The explosion vent duct from an interior dust collector is to be ducted outside through an exterior wall or up through the roof to ensure that the hot gasses are directed away from any personnel in the area or from any vulnerable equipment.

Note: An alternative to the explosion relief ductwork is the use of a device called Flamequench SQ, a flameless vent that is installed over the standard explosion vent to extinguish the flame front as it exits the vented area. Another device is the Rembe Q-Rohr indoor flameless venting system. These devices allow the venting to be inside without a duct to the outside when access to the outside is remote or venting to a safe location is not possible. These devices also connect to any audible/visual alarm and equipment shutdown to alert personnel. Consult with the manufacturer for limitations to the use of these devices, for example toxic particles are not to be expelled in the area.

An exterior dust collector must have a vent duct or blast deflector if there is personnel access to the area. In all cases, attention must be given to any access doors, ladders, or air intakes to other equipment in the area and ensure that the fireball avoids them as described in Chapter 9.

The vent duct is required to have two 8 in. long spool sections that allow the explosion vent to be replaced after an explosion occurs. The spool section farthest from the burst element is removed first, then the spool section next to the burst element can be slid away from the ruptured disk to allow replacement of the burst disk.

Any changes in angle of the duct must be in increments of 22-1/2° to keep resistance to flowing pressure to a minimum.

A filter bag cage is required to prevent the filter bags from blocking the vent area during an explosion (refer to Figure 6.13 for details).

Explosion Vented Equipment System Protection Guide, First Edition. Robert C. Comer.

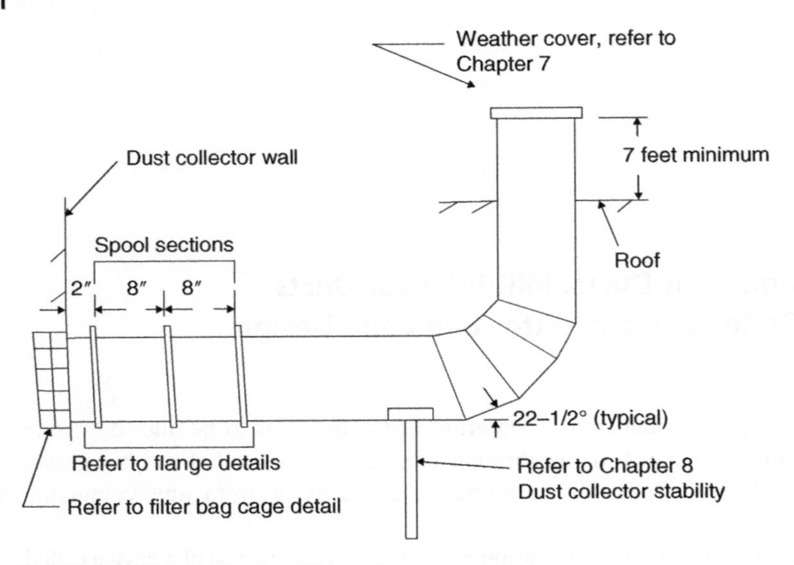

Figure 6.1 Explosion vent duct: roof.

Stencil, a warning sign near the exit plane of the explosion vent duct "(*Warning*: Stay clear, explosion Relief Device)" (Figure 6.1).

Note: The flat, horizontal, square cut weather cover on a roof can only be used where there is no snow or ice accumulation possible. The end of the duct may be cut at an angle aimed away from any personnel, equipment, or air intake in the area (Figure 6.2).

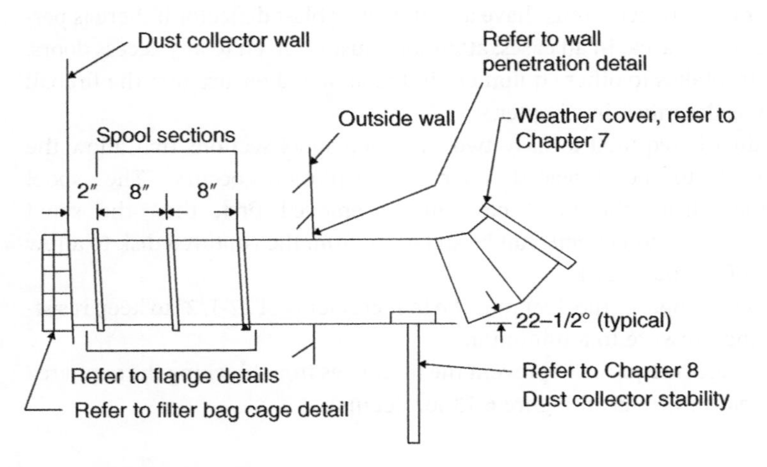

Figure 6.2 Explosion vent duct: wall.

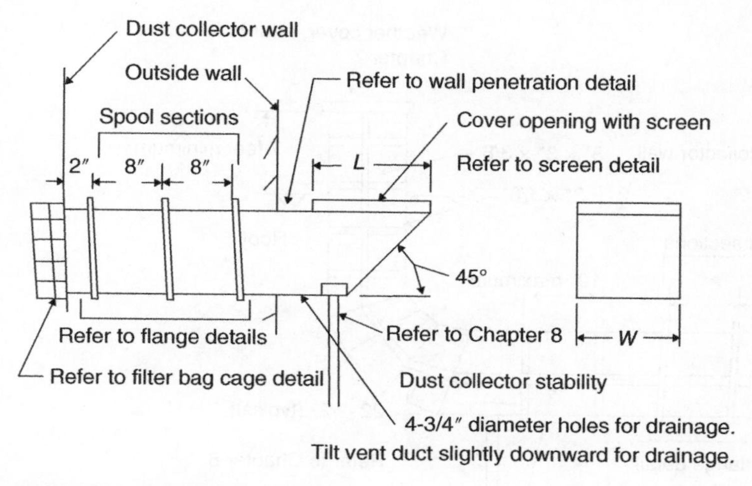

Figure 6.3 Blast deflector.

The opening in the blast deflector must have an area ($L \times W$) at least 1.2 times the vent duct flow area. The screen open area is 83.9% (see Figure 6.31); therefore, to keep resistance to the flowing pressure to a minimum, a larger opening is required (Figure 6.3).

A cylindrical vent duct very rarely requires reinforcing. A square or rectangular vent duct almost always requires reinforcing.

The requirement to reinforce sides of the vent duct or blast deflector is determined by the panel stress analyses and reinforcing calculations described in Chapter 2.

The vent duct or blast deflector should have a wall thickness of 10 ga. (0.1345″) minimum. Assume, for these examples, that the explosion vent is 24″ × 24″ (an average size that is used), then the vent duct is 24″ × 24″. The vent duct can be larger than the explosion vent if desired.

The vent duct is reinforced in the same manner as the dust collector, and since the wall thickness of the vent duct is also 10 ga. (0.1345″) for this example, the reinforcing should limit the panel sizes to less than 19″ × 19″ as previously determined (Figure 6.4).

Note: The flat, horizontal, and square cut weather cover on a roof can only be used where there is no snow or ice accumulation possible. The end of the duct may be cut at an angle aimed away from any personnel, equipment, or air intake in the area (Figure 6.5).

All four sides of the explosion vent duct are to be reinforced with 2″ × 2″ × 1/8″ angles, spaced at 19″ or less, for the short members and 3″ × 3″ × 3/8″ angles for the long members as was determined in Chapter 2.

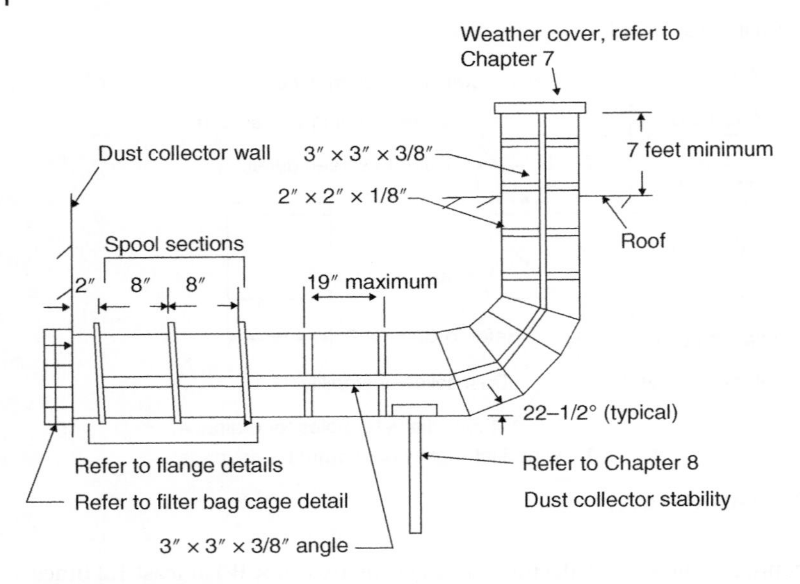

Figure 6.4 Explosion vent duct – reinforcing.

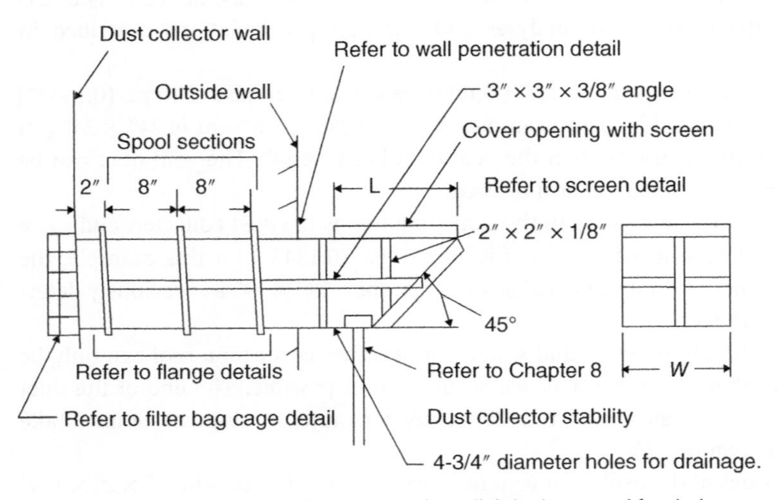

Figure 6.5 Blast deflector: reinforcing.

All four sides of the blast deflector are to be reinforced with 2″ × 2″ × 1/8″ angles, spaced at 19″ or less, for the short members and 3″ × 3″ × 3/8″ angles for the long members as was determined in Chapter 2.

The size of the blast deflector opening is to be at least 1.2 times the vent duct flow area to account for the screen blocking the area.

Then for

$$\begin{aligned} W &= 24 \text{ in.}, L = (\text{duct area}/W) \times 1.2 \\ &= (24 \times 24/24) \times 1.2 = 28.8 \text{ in., say } L = 30 \text{ in.} \end{aligned} \tag{6.1}$$

Vent Duct Flange Bolt Loading and Stress

Flange Gasket Sealing Design Requirements

The initial axial clamping load must be sufficient to keep the gasket compressed under conditions of pressure. At all times, the clamping bolt load, W_b, must be equal to or greater than the venting pressure separating load, F_r, plus the force to seal the gasket, P', plus the gasket compression force, F_g, plus the bolt force, F_w, required to support the weight of the duct. The reaction force, F_x, of the vented explosion acting on the duct will be equal to the separating load, F_r, if the duct is straight and the discharge is in line with the centerline of the duct. The weight of the duct, F_d, must be considered as acting downward and induces a bending moment in the flange with a shear load on the bolts. If the duct is not straight and is turned at the end to a discharge angle up to 90° from the centerline the reaction force will be applied at an angle to the centerline and cause an additional bending moment in the duct flange and a shear load on the bolts. The addition of a support column from the duct to the floor inside or the ground outside can reduce or eliminate the bending moment on the duct (refer to Chapter 8, dust collector stability, for column support design). For these examples no support column is considered for conservatism.

All bolts are to be the same size. The bolt load, resulting stress, and required bolt torque is based on the bolt with the highest load as determined by the following analyses. All bolts will be torqued to the same value as the highest load bolt.

The initial clamping load, W_b, is applied by specifying bolt torque. To determine bolt torque, T, the following equation is approximate and represents the best practice. The specified torque should be equal to ±10% of the value of the following equation (Figure 6.6).

$$T = 0.2 W_b D \tag{6.2}$$

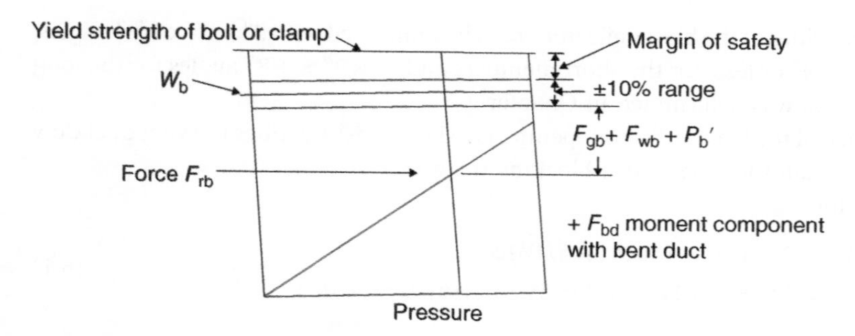

Figure 6.6 Gasketed duct flange bolt load diagram.

where

T = Bolt torque, in.-lbs
0.2 = Torque coefficient for dry steel
W_b = Axial load in bolt, lbs
D = Nominal bolt diameter, in.

where

F_r = Total venting pressure reaction force tending to separate the joint, lbs.
W_b = Individual clamping force on bolt or clamp, lbs.
F_{rb} = Individual bolt maximum reaction force, lbs.
F_{gb} = Force on bolt required to compress gasket, lbs.
P_b' = Force on bolt due to pressure on gasket required to maintain seal, lbs.
F_{wb} = Force on bolt due to weight of duct, lbs.

The minimum gasket load required to seal is the minimum sealing stress.

Assume rubber (shore $A = 65$) as a standard, economical, gasket material (no high temperature requirement because of the very short duration that the joint is exposed to the explosion venting hot gas), also assume common flat face flange joint assemblies.

The minimum sealing stress required is approximately 200 psi for firm and soft materials. A flange pressure on the seal, p', of 200 psi minimum is required to maintain a seal. The roughness of the flange face is not a factor if the minimum sealing stress is achieved. A 10–20% compression in a rubber gasket is adequate for a good gas seal.

Sealing of the flange joints in dust collectors is relatively easy with the low explosion flowing vent pressure, P_{Red}, values. The bolts are not heavily stressed, and the gasket seal is not subject to major loosening forces.

Square/rectangular vent ducts with and without bend at end of duct:

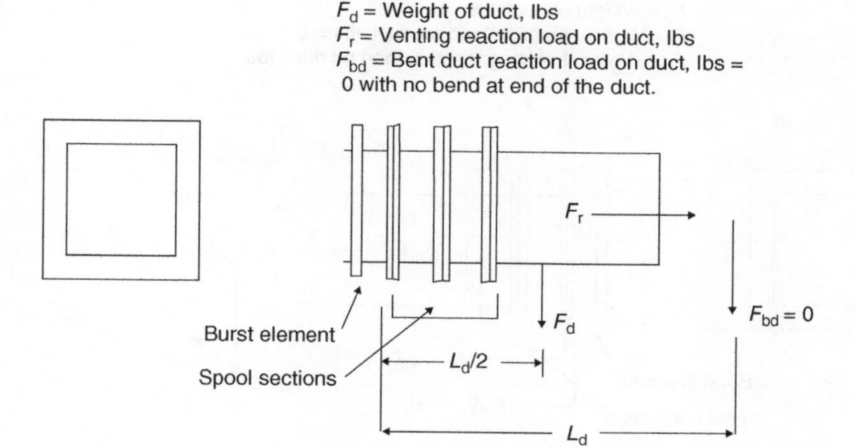

Figure 6.7 Square vent duct with no bend at end of duct.

F_d = Weight of duct, lbs
F_x = Horizontal reaction load from bent duct, lbs
= F_r (cos duct angle), lbs
F_{bd} = Bent duct vertical reaction load, lbs
= F_r (sine duct angle), lbs

Duct angle
F_x
F_{bd}
F_d
Burst element
Spool sections
$L_d/2$
L_d

Figure 6.8 Square vent duct with bend at end of duct.

Figure 6.7 illustrates the straight vent duct flange and bolt loading with no bend at end of duct ($F_{bd} = 0$).

Figure 6.8 illustrates the flange and bolt loading for square duct with a 45° bend at the end of the duct. The weight of the duct and the vertical component of the vented reaction force induces a moment load on the flange and bolts (refer to Figure 6.10 for bolt loading).

F_d = Weight of duct, lbs
F_r = Vertical reaction load on duct, lbs = 0
F_{bd} = Vertical venting reaction load on duct, lbs

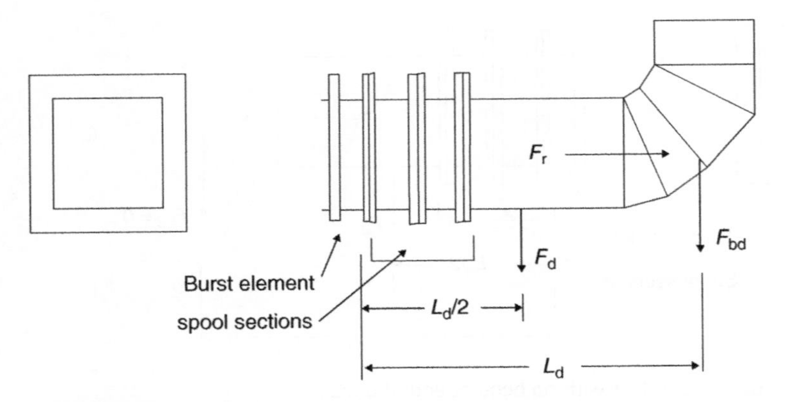

Figure 6.9 Square vent duct with 90° bend at end of duct.

Figure 6.9 illustrates the square duct flange and bolt loading with a 90° bend at the end of the duct ($F_r = 0$).

Figure 6.10 illustrates an example of a square duct flange with (24)1/2″ bolts.

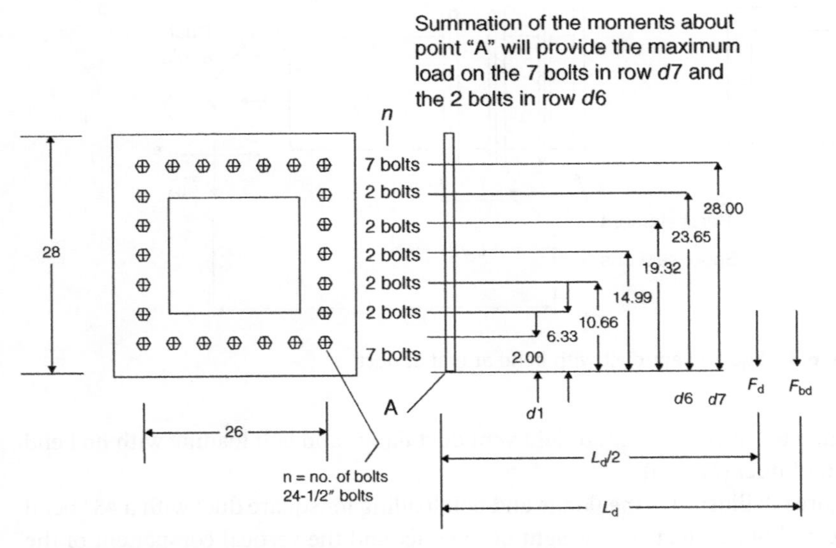

Figure 6.10 Square vent duct bolt spacing example.

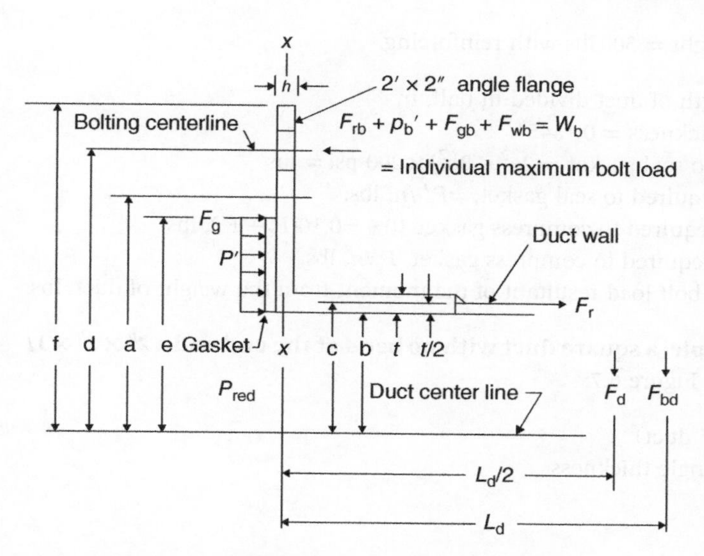

Figure 6.11 Vent duct flange detail.

Figure 6.11 illustrates in schematic form the dimensions and loading of a square straight duct with gasket loading (P'), weight of the duct loading (F_d), and bent duct loading (F_{bd}).

$$W_T = \text{Maximum total flange load} = F_r + P' + F_g + F_w \text{ lbs}$$

$$W_b = \text{Maximum individual bolt load} = F_{rb} + P_b' + F_{gb} + F_{wb} \text{ lbs}$$

F_r = Venting reaction force = a_u(DLF)(A_d)P_{Red} (refer to Chapter 8 for reaction force calculation) = 3463 lbs. *Note*: This horizontal force is for a straight vent duct with no bend on the end.

$$F_{rb} = \text{Venting reaction force per bolt, lbs} = F_r/n = 3463/24 \text{ bolts} = 144 \text{ lbs/bolt} \tag{6.3}$$

$$F_d = \text{Duct weight} = (2 \times \text{duct width} + 2 \times \text{duct height}) \times t_d \times \text{length of duct} \times 0.3 \text{ lbs/in}^3 = \text{lbs}$$
$$= 4 \times \text{side height of duct} \times t_d \times \text{length of duct} \times 0.3 \text{ lbs/in}^3$$
$$= 4 \times 24/\text{side} \times 0.1345 \times 120 \times 0.3 \times 465 \text{ lbs} \tag{6.4}$$

L_d = Length of horizontal duct
L_{dext} = Length of extended duct bended section
F_d = Approximate weight of the square duct

Assume duct weight = 500 lbs with reinforcing.

$L_d/2$ = Total length of duct divided in half, in.
t_d = Duct wall thickness = 0.1345″.
P' = Total force to seal gasket $=(2e^2 - 2b^2) \times 200$ psi = lbs
P_b = Bolt force required to seal gasket, $=P'/n$, lbs.
F_g = Total load required to compress gasket 10% $=0.10(F_r + P')$, lbs.
F_{gb} = Bolt force required to compress gasket, F_g/n, lbs.
F_{wb} = Maximum bolt load resultant of the moment from the weight of duct, lbs.

For this example, a square duct with no bend at the end and a 2″ × 2″ × 1/4″ angle flange, Figure 6.7:

$b = 12''$ (for a 24″ duct)
$h = 1/4''$ flange angle thickness
$h/2 = 0.125''$
$c = 12.125''$
$f = 14''$
$d = 13.00''$
$e = 12.75''$ (gasket width = 0.75″)
$n = 24$ bolts
L_d = Length of duct, in. = 120 in.
$L_{dext} = 0$ in.
$L_d/2$ = Distance that weight of duct acts on flange = 60 in.

Then,

$$P' = (25.5^2 - 24^2)200 = 14\,850 \text{ lbs}$$

$$P_b' = P'/n = 14\,850/24 = 618 \text{ lbs/bolt} \tag{6.5}$$

$$F_g = 0.10(3463 + 14\,850) = 1831 \text{ lbs}$$

$$F_{gb} = F_g/n = 1831/24 = 76 \text{ lbs/bolt} \tag{6.6}$$

Solving for the load from the moment induced by the duct weight, F_d: refer to Figure 6.10. The total bolt load, W_T, is equally distributed over the 24 bolts when the duct is straight and the weight of the duct is not included in the calculations. When the weight of the duct is included and the duct is bent, the duct bent force is directed at an angle and induces a moment in the flange, a shear load on the bolts, and the bolt load is not equally distributed. Refer to the following examples.

Assume that the vent and duct is 24″ × 24″ and the flange is 2″ × 2″ × 1/4 angle.

Assume that there is no column support for a conservative analysis. The load is taken entirely by the flange and bolts.

The duct flange bolt arrangement is to match the explosion vent flange. For this example assume that there are (24)1/4″ bolts (refer to Figure 6.10).

All duct flanges are matched to the explosion vent flange. The bolts attaching the first duct flange on the spool section to the vent flange are the most highly loaded and stressed bolts.

Combined shear and tension in the 1/2″ bolts: the maximum bolt loading is on the seven bolts farthest from point "A" at the top, distance d7, and the two bolts at distance d6. Summation of the moments about point "A" will provide the maximum load on the 7 bolts in row d7 and the 2 bolts in row d6.

Where:

A_r = Stress area of 1/2″ bolt = 0.1597 in.^{2}

A_d = Full diameter shear area at 0.4387 minor diameter = 0.1512 in.^{2}

n = Number of bolts = 24

M = Venting load moment on bent duct = F_{bd} × length of duct, in.-lbs for this example = 0 (no bent duct)

$$\begin{aligned} M_D &= \text{duct weight load moment} = F_d \times 1/2 \text{ length of duct} \\ &= 500 \times 60 = 30\,000 \text{ in.-lbs} \end{aligned} \tag{6.7}$$

$$f' = 1/2\left(f_t + \sqrt{f_t^2 + 4f_s^2}\right) = \text{combined bolt tension and shear stress, psi} \tag{6.8}$$

$$f_s = (F_{bd} + F_d)/nA_d = 0 + 500/(24 \times 0.1512) = 138 \text{ psi bolt shear stress} \tag{6.9}$$

$$f_s = F_b/A_r = \text{total axial bolt tension stress due to moment, psi}$$

For the 7 bolts in Top Row, Distance d7 from Point "A":

$$F_{wb} = (F_{bd} \times L_d + F_d \times L_d/2)(d7)/\Sigma(n_1)(d_1^2) + (n_2)(d_2^2) + \cdots + (n_7)(d_7^2) \tag{6.10}$$

$$\begin{aligned} F_{wb} &= (0 + 500 \times 60)(28)/(7)(2.0^2) + (2.0)(6.33^2 + 10.66^2 + 14.99^2 + 19.32^2 + 23.65^2) + (7)(28^2) \\ &= 840\,000/8143 = 103 \text{ lbs/7 bolts axial tension load maximum} \end{aligned} \tag{6.11}$$

The load per bolt in row d7 is 103/7 = 15 lbs/bolt.

The duct weight bolt loading is not very high.

$$W_b = F_{rb} + P_b' + F_{gb} + F_{wb} \tag{6.12}$$

$$W_b = 144 + 618 + 76 + 15 = 853 \text{ lbs/bolt} \tag{6.13}$$

$$f_t = W_b/A_r = 853/0.1597 = 5341 \text{ psi/bolt} \tag{6.14}$$

Then,

$$f' = 1/2\left(5341 + \sqrt{5341^2 + 4 \times 138^2}\right)$$
$$= 1/2(5341 + 5348) = 5345 \text{ psi combined bolt tension and shear stress, psi} \quad (6.15)$$

Refer to Table 5.1 for allowable bolt stress.

$$\text{The factor of safety} = 24120/5345 = 4.5 \quad (6.16)$$
$$\text{The required bolt torque is} = 0.2 \times 853 \times 0.5 = 85 \text{ in.-lbs} \quad (6.17)$$

Check on the two bolts on row 23.65″, distance d6 from the bottom point "A":

$$f' = 1/2\left(f_t + \sqrt{f_t^2 + 4f_s^2}\right)$$
$$f_s = F_{bd} + F_D/nA_d = (0 + 500)/24 \times 0.1512 = 138 \text{ psi shear stress} \quad (6.18)$$

All of the factors remain the same except that d7 is replaced by d6.

$$f_t = F_b/A_r$$
$$F_{wb} = (F_{bd} \times L_d + F_d \times L_d/2)(d6)/\Sigma(n_1)(d_1^2) + (n_2)(d_2^2) + \cdots + (n_7)(d_7^2) \quad (6.19)$$

$$F_{wb} = (0 + 500 \times 60)(23.65)/8143 = 709500/8143 = 87 \text{ lbs/2 bolts in row d6}$$
Load per bolt in row d6 is $87/2 = 43.5$ lbs/bolt. (6.20)

$$W_b = 144 + 618 + 76 + 43.5 = 882 \text{ lbs/bolt} \quad (6.21)$$

Then,

$$f_t = W_b/A_r = 882/0.1597 = 5520 \text{ psi} \quad (6.22)$$

Then,

$$f' = 1/2\left(5520 + \sqrt{5520^2 + 4 \times 138^2}\right) = 1/2(5520 + 5526)$$
$$= 5523 \text{ psi combined bolt tension and shear stress, psi} \quad (6.23)$$

Refer to Table 5.1 for allowable bolt stress.

$$\text{The factor of safety} = 24\,120/5523 = 4.4 \quad (6.24)$$
$$\text{The required bolt torque is } 0.2 \times 882 \times 0.5 = 88 \text{ in.-lbs} \quad (6.25)$$

The two bolts having the highest loading determine that the required torque on all bolts is 88 in.-lbs.

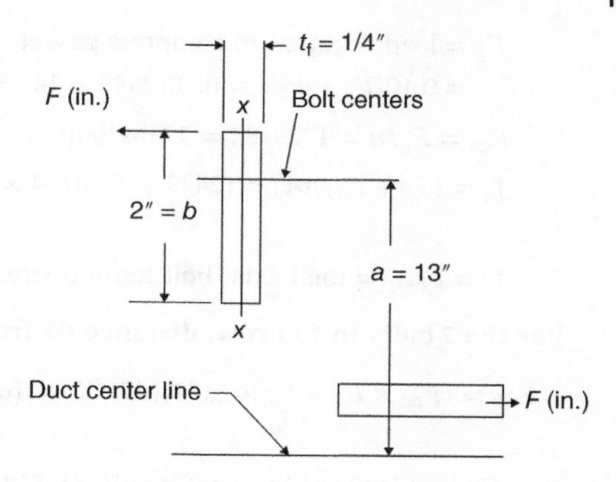

Figure 6.12 Flange stress with no bend.

Figure 6.12 Flange stress with no bend at end of duct (for the square bolted flange, refer Figure 6.10).

The worst bolt load is 882 lbs on the two bolts from the case where there is no bend at the end of the duct.

$$\text{The span between bolts is} = 4 \times 26/24 \text{ bolts} = 4.33 \text{ in.} \tag{6.26}$$

Then,

$$F/\text{in.} = \text{the load on the flange} = 882/4.33 = 204 \text{ lbs/in.} \tag{6.27}$$

$$I_{x-x} = bt_f^3/12 = 2(0.25^3)/12 = 0.0026 \text{ in.}^4 \tag{6.28}$$

$$M_t = F/\text{in.} \times b = 204 \times 2 = 408 \text{ in.-lbs/in.} \tag{6.29}$$

$$C = t_f/2 = 0.25/2 = 0.125 \text{ in.} \tag{6.30}$$

$$\text{Flange stress} = M_t C/I_{x-x} = 408(0.125)/0.0026 = 19\,615 \text{ psi} \tag{6.31}$$

Factor of safety = 24 120/19 615 = 1.23 (the flange is safe).

Square duct with a 45° bend at the end of the duct (refer to Figure 6.8, bolted flange):

Where there is a bent duct condition as shown in Figure 6.8, there is an additional moment due to the deflected venting pressure equal to $F_{bd} \times L_d$, and the horizontal axial load is reduced from F_r to F_x. For the duct angle equal to 45°

$$F_{bd} = F_r \times \sin 45° = 3463 \times 0.7071 = 2449 \text{ lbs}$$

and (6.32)

$$F_x = F_r \times \cos 45° = 3463 \times 0.7071 = 2449 \text{ lbs}$$

F_{xb} = venting reaction force per bolt

$= F_x/n = 2449/24 \text{ bolts} = 102 \text{ lbs/bolt}$

$$F_g = \text{load required to compress gasket} = 0.10(F_x + P_b') = 0.10(2449 + 14\,850) = 1729 \text{ lbs} \quad (6.33)$$

$$F_{gb} = F_g/n = 1729/24 = 72 \text{ lbs/bolt} \quad (6.34)$$

$$f_s = F_{bd} + F_d/nA_d = (2449 + 500)/24 \times 0.1512 = 813 \text{ psi bolt shear stress} \quad (6.35)$$

$$f_t = F_b A_r = \text{total axial bolt tension stress due to moment, psi} \quad (6.36)$$

For the 7 bolts in top row, distance d7 from point "A":

$$F_{wb} = (F_{bd} \times L_d + F_d \times L_d/2)(d7)/\Sigma(n_1)(d_1^2) + (n_2)(d_2^2) + \cdots + (n_7)(d_7^2) \quad (6.37)$$

$$F_{wb} = (2449 \times 120 + 500 \times 60)(28)/8143 = 9\,068\,640/8143 = 1114 \text{ lbs/7 bolts axial tension load maximum} \quad (6.38)$$

Load per bolt in row d7 is 1114/7 = 159 lbs/bolt

$W_b = 102 + 616 + 72 + 159 = 951$ lbs/bolt

$F_t = W_b/A_r = 951/0.1597 = 5955$ psi/bolt

Then,

$$f' = 1/2\left(5955 + \sqrt{5955^2 + 4 \times 138^2}\right) = 1/2(5955 + 5961) = 5958 \text{ psi combined bolt tension and shear stress, psi} \quad (6.39)$$

Refer to Table 5.1 for allowable bolt stress.

The factor of safety = 24 120/5958 = 4.0 (6.40)

The required bolt torque is 0.2 × 951 × 0.5 = 95 in.-lbs (6.41)

Check on the two bolts on row 23.65″, distance d6 from the bottom point "A": All of the factors remain the same except d7 is replaced by d6.

$$f' = 1/2\left(f_t + \sqrt{f_t^2 + 4f_s^2}\right)$$

$$f_s = F_{bd} + F_D/nA_d = (2449 + 500)/24 \times 0.1512 = 813 \text{ psi shear stress} \quad (6.42)$$

All of the factors remain the same except that d7 is replaced by d6.

$$f_t = F_b/A_r$$

$$F_{wb} = (F_{bd} \times L_d + F_d \times L_d/2)(d6)/\Sigma(n_1)(d_1^2) + (n_2)(d_2^2) + \cdots + (n_7)(d_7^2) \quad (6.43)$$

$$F_{wb} = (2449 \times 120 + 500 \times 60)(23.65)/8143 = 7\,659\,762/8143$$
$$= 941 \text{ lbs/2 bolts in row d6} \quad (6.44)$$

Load per bolt in row d6 is 941/2 = 470 lbs/bolt

$$W_b = 102 + 618 + 72 + 470 = 1262 \text{ lbs/bolt} \quad (6.45)$$

Then,

$$f_s = 1262/0.1597 = 7902 \text{ psi}$$

Then,

$$f' = 1/2\left(7902 + \sqrt{7902^2 + 4 \times 813^2}\right) = 1/2(7902 + 8129)$$
$$= 8015 \text{ psi combined bolt tension and shear stress,} \quad (6.46)$$

$$\text{The factor of safety} = 24\,120/8015 = 3.0 \quad (6.47)$$

The required bolt torque is 0.2 × 1262 × 0.5 = 126 in.- lbs. The torque for the two bolts (126 in.-lbs) is the required torque (Figure 6.13).

The worst bolt load is 1262 lbs on the two bolts from the case where there is a 45° bend at the end of the duct.

$$\text{The span between bolts is} = 4 \times 26/24 \text{ bolts} = 4.33 \text{ in.} \quad (6.48)$$

Then,

$$F(\text{in.}) = \text{the load on the flange} = 1262/4.33 = 291 \text{ lbs/in.} \quad (6.49)$$
$$I_{x-x} = bt_f^3/12 = 2(0.25^3)/12 = 0.0026 \text{ in.}^4 \quad (6.50)$$
$$M_t = F/\text{inch} \times b = 291 \times 2 = 582 \text{ in.-lbs/in.} \quad (6.51)$$
$$C = t_f/2 = 0.25/2 = 0.125 \text{ in.} \quad (6.52)$$
$$\text{Flange stress} = M_t C/I_{x-x} = 582(0.125)/0.0026 = 27\,980 \text{ psi} \quad (6.53)$$

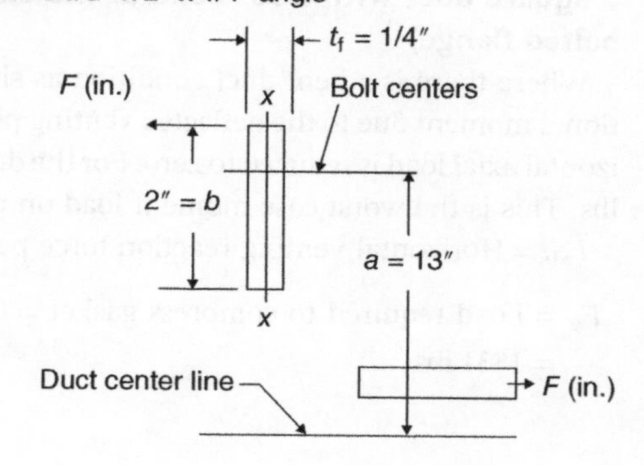

Figure 6.13 Flange stress with a 45° bend at the end of the duct (for the square bolted flange, refer Figure 6.10).

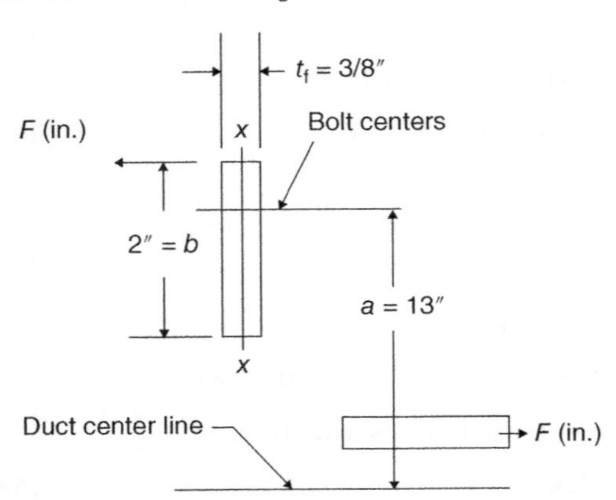

Figure 6.14 Flange stress with a 45° bend (for the square bolted flange, refer Figure 6.10).

Factor of safety = 24 120/27 980 = 0.862 (the flange is overstressed).

Try increasing flange thickness to 3/8″ (Figure 6.14).

The worst bolt load is 1262 lbs on the two bolts from the case where there is a 90° bend at the end of the duct.

$$\text{The span between bolts is} = 4 \times 26/24\ \text{bolts} = 4.33\ \text{in.} \tag{6.54}$$

Then,

$$F/\text{inch} = \text{The load on the flange} = 1262/4.33 = 291\ \text{lbs/in.} \tag{6.55}$$

$$\vec{I}_{x-x} = bt_f^{\,3}/12 = 2(0.375^3)/12 = 0.0088\ \text{in.}^4 \tag{6.56}$$

$$M_t = F/\text{inch} \times b = 291 \times 2 = 582\ \text{in.-lbs/in.} \tag{6.57}$$

$$C = t_f/2 = 0.375/2 = 0.188\ \text{in.} \tag{6.58}$$

$$\text{Flange stress} = M_t C/I_{x-x} = 582(0.188)/0.0088 = 12\,434\ \text{psi} \tag{6.59}$$

$$\text{Factor of safety} = 24\,120/12\,434 = 1.94\ \text{(the 3/8″flange is safe)} \tag{6.60}$$

Square duct with a 90° bend at the end of the duct (refer to Figure 6.9, **bolted flange):**

Where there is a bent duct condition as shown in Figure 6.9, there is an additional moment due to the deflected venting pressure equal to $F_{bd} \times L_d$ and the horizontal axial load is reduced to zero. For the duct angle equal to 90° $F_{bd} = F_r = 3463$ lbs. This is the worst case moment load on the flange.

F_{xb} = Horizontal venting reaction force per bolt, lbs = 0 lbs/bolt

$$F_g = \text{Load required to compress gasket} = 0.10(F_x + P_b') = 0.10(3463 + 14\,850) = 1831\ \text{lbs} \tag{6.61}$$

$$F_{gb} = F_g/n = 1831/24 = 76 \text{ lbs/bolt} \quad (6.62)$$

$$f_s = F_{bd} + F_d/nA_d = (3463 + 500)/24 \times 0.1512 = 1092 \text{ psi bolt shear stress} \quad (6.63)$$

$f_t = F_b/A_r =$ Total axial bolt tension stress due to moment, psi

$$F_d = \text{Duct weight} = (2 \times \text{duct width} + 2 \times \text{duct height}) \times t_d \times (L_d + L_{dext}) \times 0.3 \text{ lbs/in.}^3 = \text{lbs}$$
$$= 4 \times \text{side height of duct} \times t_d \times (L_d + L_{dext}) \times 0.3 \text{ lbs/in.}^3$$
$$= 4 \times 24/\text{side} \times 0.1345 \times (120 + 30) \times 0.3 = 581 \text{ lbs} \quad (6.64)$$

where

L_d = Length of horizontal duct
L_{dext} = Length of extended duct bended section
F_d = Approximate weight of the square duct
Assume duct weight = 650 lbs with reinforcing.

For the 7 bolts in top row, distance d7 from point "A":

$$F_{wb} = (F_{bd} \times L_d + F_d \times L_d/2)(d7)/\Sigma(n_1)(d_1^2) + (n_2)(d_2^2) + \cdots + (n_7)(d_7^2) \quad (6.65)$$

$$F_{wb} = (3463 \times 120 + 650 \times 60)(28)/8143 = 12727680/8143$$
$$= 1563 \text{ lbs/7 bolts axial tension load max.}$$

Load per bolt in row d7 is 1563/7 = 223 lbs/bolt

$$W_b = 0 + 618 + 76 + 223 = 917 \text{ lbs/bolt} \quad (6.66)$$

$$F_t = W_b/A_r = 917/0.1597 = 5742 \text{ psi/bolt} \quad (6.67)$$

Then,

$$f' = 1/2\left(5742 + \sqrt{5742^2 + 4 \times 1092^2}\right) = 1/2(5742 + 6143)$$
$$= 5942 \text{ psi combined bolt tension and shear stress} \quad (6.68)$$

Refer to Table 5.1 for allowable bolt stress.

The factor of safety = 24 120/5942 = 4.0 (6.69)

The required bolt torque is 0.2 × 917 × 0.5 = 92 in.-lbs (6.70)

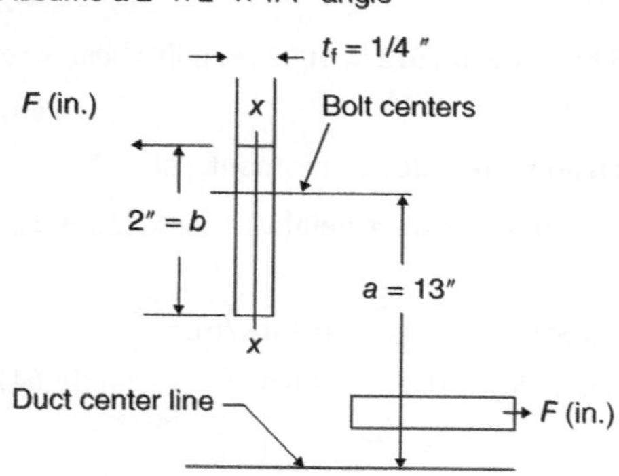

Figure 6.15 Flange stress with a 90° bend at the end of the duct (for the square bolted flange, refer Figure 6.10).

Check on the two bolts on row 23.65″, distance d6 from the bottom point "A": All of the factors remain the same except d7 is replaced by d6 (Figure 6.15).

$$f' = 1/2\left(f_t + \sqrt{f_t^2 + 4f_s^2}\right) \quad (6.71)$$

$$f_t = F_b/A_r \quad (6.72)$$

$$F_{wb} = (F_{bd} \times L_d + F_d \times L_d/2)(d6)/\Sigma(n_1)(d_1^2) + (n_2)(d_2^2) + \cdots + (n_7)(d_7^2) \quad (6.73)$$

$$F_{wb} = (3463 \times 120 + 650 \times 60)(23.65)/8143 = 10\,750\,344/8143$$
$$= 1320 \text{ lbs/2 bolts axial tension load max.} \quad (6.74)$$

Load per bolt in row d6 is 1320/2 = 660 lbs/bolt

$$W_b = F_{rb} + p_b' + F_{gb} + F_{wb} \quad (6.75)$$

$$W_b = 0 + 618 + 76 + 660 = 1354 \text{ lbs/bolt} \quad (6.76)$$

Then,

$$f_t = 1354/0.1597 = 8478 \text{ psi} \quad (6.77)$$

Then,

$$f' = 1/2\left(8478 + \sqrt{8478^2 + 4 \times 1092^2}\right) = 1/2(8478 + 8754)$$
$$= 8616 \text{ psi combined bolt tension and shear stress} \quad (6.78)$$

The factor of safety = 24120/8616 = 2.8 (6.79)

The required bolt torque is = 0.2 × 1354 × 0.5 = 135 in.-lbs. (6.80)

The torque for the two bolts (135 in.-lbs) is the required torque.

The worst bolt load is 1354 lbs on the two bolts from the case where there is a 90° bend at the end of the duct.

The span between bolts is = 4 × 26/24 bolts = 4.33 in. (6.81)

Then,

$$F/\text{inch} = \text{the load on the flange} = 1354/4.33 = 312\ \text{lbs/in.} \quad (6.82)$$

$$I_{x-x} = bt_f^3/12 = 2(0.25^3)/12 = 0.0026\ \text{in.}^4 \quad (6.83)$$

$$M_t = F/\text{inch} \times b = 312 \times 2 = 624\ \text{in.-lbs/in.} \quad (6.84)$$

$$C = t_f/2 = 0.25/2 = 0.125\ \text{in.} \quad (6.85)$$

$$\text{Flange stress} = M_t C/I_{x-x} = 624(0.125)/0.0026 = 30\,000\ \text{psi} \quad (6.86)$$

Factor of safety = 24 120/30 000 = 0.804 (the flange is overstressed).
Try increasing flange thickness to 3/8″ (Figure 6.16).

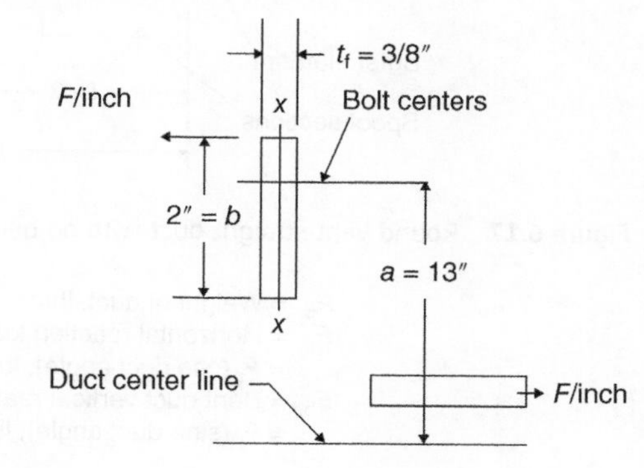

Figure 6.16 Flange stress with a 90° bend (for the square bolted flange, refer Figure 6.10).

The worst bolt load is 1354 lbs on the two bolts from the case where there is a 90° bend at the end of the duct.

$$\text{The span between bolts is} = 4 \times 26/24\ \text{bolts} = 4.33\ \text{in.} \quad (6.87)$$

Then,

$$F/\text{inch} = \text{the load on the flange} = 1354/4.33 = 312\ \text{lbs/in.} \quad (6.88)$$

$$I_{x-x} = bt_f^3/12 = 2(0.375^3)/12 = 0.0088\ \text{in.}^4 \quad (6.89)$$

$$M_t = F/\text{inch} \times b = 312 \times 2 = 624\ \text{in.-lbs/in.} \quad (6.90)$$

$$C = t_f/2 = 0.375/2 = 0.188\ \text{in.} \quad (6.91)$$

$$\text{Flange stress} = M_t C/I_{x-x} = 624(0.188)/0.0088 = 13330\ \text{psi} \quad (6.92)$$

$$\text{Factor of safety} = 24\,120/13\,330 = 1.81\ (\text{the 3/8″flange is safe}) \quad (6.93)$$

For a circular bolted flange for a straight duct, no bend at end with a 2″ × 2″ × 1/4″ angle flange (Figures 6.17–6.20):

F_d = Weight of duct, lbs
F_x = Horizontal reaction load from bent duct, lbs
= F_r (cos duct angle), lbs
F_{bd} = Bent duct vertical reaction load, lbs
= F_r (sine duct angle), lbs

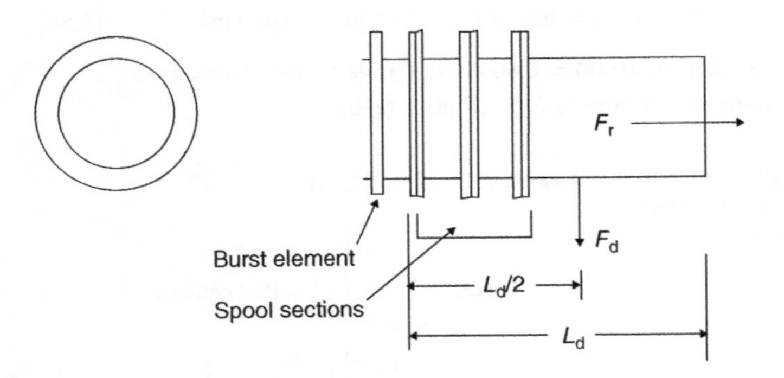

Figure 6.17 Round vent straight duct with no bend at end of duct.

F_d = Weight of duct, lbs
F_x = Horizontal reaction load from bent duct, lbs
= F_r (cos duct angle), lbs
F_{bd} = Bent duct vertical reaction load, lbs
= F_r (sine duct angle), lbs

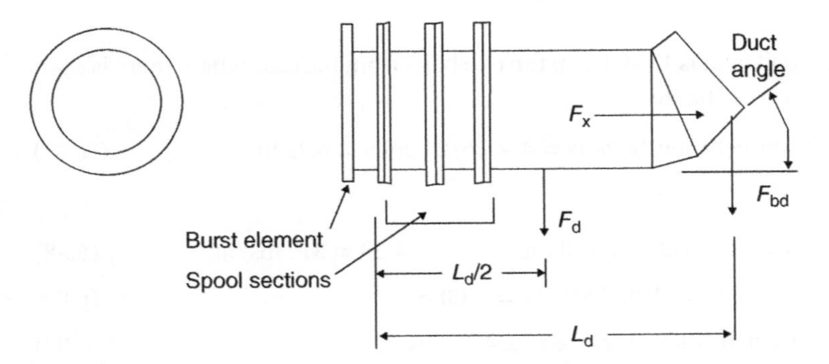

Figure 6.18 Round vent straight duct with 45° bend at end of duct.

$$W_T = \text{Maximum total flange load} = F_r + P' + F_g + F_w, \text{lbs} \tag{6.94}$$

$$W_b = \text{Maximum individual bolt load} = F_{rb} + P_b{}' + F_{gb} + F_{wb}, \text{lbs} \tag{6.95}$$

$$F_r = \text{Venting reaction force} = a_u(\text{DLF})(A_d)P_{\text{Red}}, \text{lbs} \tag{6.96}$$

(Refer to Chapter 8 for reaction force calculation) = 3463 lbs. *Note*: This horizontal force is for a straight vent duct with no bend on the end.

F_d = Weight of duct, lbs
F_x = Horizontal venting load on duct = 0 lbs
F_{bd} = Vertical venting reaction load on duct, lbs

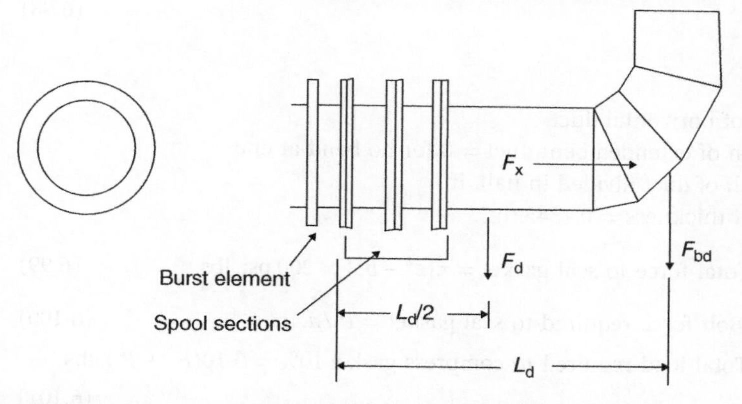

Figure 6.19 Round vent straight duct with 90° bend at end of duct.

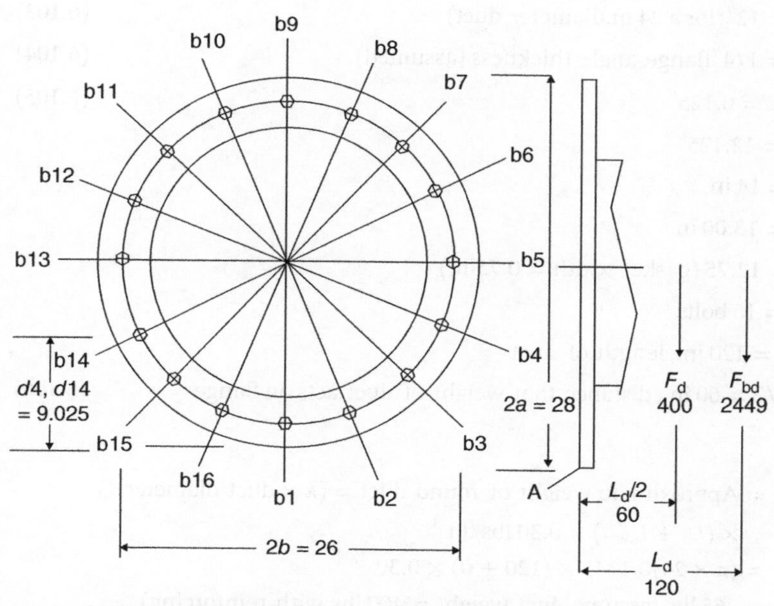

Figure 6.20 Round vent duct bolted flange 16 blots example.

$$F_{rb} = F_r/n = 3463/16 \text{ bolts} = 216 \text{ lbs/bolt} \tag{6.97}$$

$$F_d = \text{Duct weight} = (\pi \times \text{duct diameter}) \times t_d \times (L_d + L_{dext}) \times 0.3 \text{ lbs/in.}^3, \text{lbs} \tag{6.98}$$

where

L_d = Length of horizontal duct.
L_{dext} = Length of extended bent duct = 0 for no bend at end.
$L_d/2$ = Length of duct divided in half, in.
t_d = duct wall thickness = 0.1345, in.

$$P' = \text{Total force to seal gasket} = \pi\left(e^2 - b^2\right) \times 200 \text{ psi, lbs} \tag{6.99}$$

$$P_b{}' = \text{Bolt force required to seal gasket} = P'/n, \text{lbs} \tag{6.100}$$

$$F_g = \text{Total load required to compress gasket } 10\% = 0.10(F_r + P'), \text{lbs} \tag{6.101}$$

$$F_{gb} = \text{Bolt force required to compress gasket} = F_g/n, \text{lbs} \tag{6.102}$$

F_{wb} = Maximum bolt load resultant of the moment from the weight of duct, lbs.

$$b = 12''(\text{for a 24 in.diameter duct}) \tag{6.103}$$

$$h = 1/4''\text{flange angle thickness (assumed)} \tag{6.104}$$

$$h/2 = 0.125'' \tag{6.105}$$

$$C = 12.125''$$

$$f = 14 \text{ in.}$$

$$d = 13.00 \text{ in.}$$

$$e = 12.75 \text{ (gasket width} = 0.75\text{in.)}$$

$$n = 16 \text{ bolts}$$

$$L_d = 120 \text{ in., length of duct}$$

$$L_d/2 = 60 \text{ in., distance that weight of duct acts on flange} \tag{6.106}$$

Then,

$$\begin{aligned} F_d &= \text{Approximate weight of round duct} = (\pi \times \text{duct diameter})t_d \\ &\quad \times (L_d + L_{dext}) \times 0.30 \text{ lbs/in.}^3 \\ &= (\pi \times 24)0.1345 \times (120 + 0) \times 0.30 \\ &= 365 \text{ lbs (assume duct weight} = 400 \text{ lbs with reinforcing)} \end{aligned}$$

$$P' = \pi\left(12.75^2 - 12^2\right)200 = 11\,663 \text{ lbs} \tag{6.107}$$

$$P_b{}' = P'/n = 11\,663/16 = 729 \text{ lbs/bolt} \tag{6.108}$$

$$F_g = 0.10(3463 + 11\,663) = 1513 \text{ lbs} \tag{6.109}$$

$$F_{gb} = F_g/n = 1513/16 = 95 \text{ lbs/bolt} \tag{6.110}$$

Solving for the load from the moment induced by the duct weight, F_d: refer to Figures 6.17 and 6.20. The total bolt load, W_T, is equally distributed over the 16 bolts when the duct is straight, and the weight of the duct is not included in the calculations. When the weight of the duct is included and the duct is bent, the duct bent force is directed at an angle and induces a moment in the flange, a shear load on the bolts and the bolt load is not equally distributed. Refer to the following examples.

Assume that the vent and duct is 24″ diameter. The flange is 2″ × 2″ × 1/4″ angle.

Assume that there is no column support for a conservative analysis. The entire load is taken by the flanges and bolts.

The duct flange bolt arrangement is to match the explosion vent flange. For this example assume that there are (16)1/2″ bolts (refer to Figure 6.20).

All duct flanges are matched to the explosion vent flange. The bolts attaching the first duct flange on the spool section to the vent flange are the most highly loaded and stressed bolts.

Combined shear and tension in the 1/2″ bolts: the maximum bolt loading is on the b9 bolt farthest from point "A" at the top, distance d9. The summation of the moments about point "A" will provide the maximum load on the b9 bolt in distance d9.

Where:

A_r = Stress area of 1/2 bolt = 0.1597 in.^{2}

A_d = Full diameter shear area at 0.4387 minor diameter = 0.1512 in.^{2}

n = Number of bolts = 16

M = Venting load moment on bent duct = $F_{bd} \times L_d/2$, in.-lbs.
For this example = 0 (no bent duct) (6.111)

M_D = Duct weight load moment = $F_d \times L_d/2 = 400 \times 60 = 24\,000$ in.-lbs (6.112)

$$f' = 1/2\left(f_t + \sqrt{f_t^2 + 4f_s^2}\right) = \text{Combined bolt tension and shear stress, psi} \tag{6.113}$$

$$f_s = F_{bd} + F_d/nA_d = 0 + 400/(16 \times 0.1512) = 165 \text{ psi bolt shear stress} \tag{6.114}$$

$$f_t = F_b/A_r = \text{Total axial bolt tension stress due to the moment, psi} \tag{6.115}$$

The load for the bolt b9 at the distance d9 from point "A":

Summation of the moments about point "A":

$$F_{bd}L_d/2 = (Fb9/d9)\left(d1^2 + d2^2 + d3^2 + \cdots + d9^2\right), \quad (6.116)$$
where, $d1 = a - b\cos\alpha$, etc.

By substituting values of d into the foregoing equation:

$$F_{b1} = (F_{bd}L_d/2)(a - b\cos\alpha)/\left(na^2 + nb^2/2\right) \quad (6.117)$$

The maximum tension load, F_{bmax}, in any bolt will occur at a distance $a + b$ from A.

This gives the general equation:

$$F_{bmax} = (F_{bd}L_d/2)(a + b)/na^2 + nb^2/2 \quad (6.118)$$

where n = number of equal size, equally spaced bolts.

$$\begin{aligned}F_{bmax} &= 400 \times 60(14 + 13)/16 \times 14^2 + 16 \times 13^2/2 = 648\,000/(3136 + 1352)\\ &= 648\,000/4488 = 144 \text{ lbs occuring in bolt b9}\end{aligned} \quad (6.119)$$

Combined shear and tension in the 1/2″ bolts: the maximum bolt loading is on the bolt "b9" farthest from point "A" at the top.

The shear stress is equal on all of the bolts and is:

$$f_s = F_{bd}/nA_d$$

where:

A_r = Stress area of 1/2 bolt = 0.1597 in.^{2}

A_d = Full diameter shear area at 0.4387 minor diameter = 0.1512 in.^{2}

n = Number of bolts = 16

$$f_s = F_{bd} + F_d/nA_d = 0 + 400/(16 \times 0.1512) = 165 \text{ psi bolt shear stress} \quad (6.120)$$

and,

$$f_t = F_{bmax}/A_r = 144/0.1597 = 902 \text{ psi} \quad (6.121)$$

Combining stresses:

$$f' = 1/2\left(f_t + \sqrt{f_t^2 + 4f_s^2}\right) \quad (6.122)$$

Then,

$$f' = 1/2\left(902 + \sqrt{902^2 + 4 \times 165^2}\right) = 1/2(902 + 960) = 931 \text{ psi} \quad (6.123)$$

The factor of safety = 24 120/931 = 26 (6.124)

W_b = Maximum individual bolt load = $F_{rb} + P_b' + F_{gb} + F_{wb}$
= 216 + 729 + 95 + 144 = 1184 lbs (6.125)

The required bolt torque is = 0.2 × 1184 × 0.5 = 118 in.-lbs (6.126)

For a circular bolted flange with a 45° bend at the end of the duct and a 2″ × 2″ × 1/4″ angle flange (refer to Figures 6.18 and 6.20**):**

All of the conditions are the same except that F_r becomes F_x and F_{bd} is acting at end of duct

W_T = Maximum total flange load = $F_r + P' + F_g + F_w$, lbs (6.127)

W_b = Maximum individual bolt load = $F_{rb} + P_b' + F_{gb} + F_{wb}$, lbs (6.128)

F_r = Venting reaction force = $a_u(\text{DLF})(A_d)P_{\text{Red}}$ (refer to Chapter 8 for reaction force calculation) = 3463 lbs. *Note*: This horizontal force is for a straight vent duct with no bend on the end.

Assume a 45° angle at the end of the duct (refer to Figure 6.17).

F_x = Horizontal component of the reaction force with a bend at the end of the duct, lbs (6.129)
= F_r sin duct angle = 3463 × 0.7071 = 2449 lbs

F_y = Vertical component of the reaction force with a bend at the end of the duct, lbs (6.130)
= F_r cosine duct angle = 3463 × 0.7071 = 2449 lbs

F_{rb} = Venting reaction force per bolt, lbs
= F_r/n = 2449/16 bolts = 153 lbs/bolt (6.131)

F_d = Duct weight = 400 lbs from previous example

t_d = Duct wall thickness = 0.1345 in.

P' = Total force to seal gasket = $\pi(e^2 - b^2) \times 200$ psi
= $\pi(12.75^2 - 12^2) \times 200$ = 11 663 lbs (6.132)

P_b' = Bolt force required to seal gasket = P'/n = 11 663/16 = 729 lbs/bolt (6.133)

F_g = Total load required to compress gasket 10% = $0.10(F_x + P')$
= 0.10(2449 + 11 663) = 1411 lbs (6.134)

$$F_{gb} = \text{Bolt force required to compress gasket} = F_g/n = 1411/16 = 88\text{lbs/bolt} \quad (6.135)$$

F_{wb} = Maximum bolt load resultant of the moment from the weight of duct, lbs.
F_{bd} = Maximum bolt load resultant of the moment from the bent duct, lbs.
$b = 12''$ (for a 24″ duct)
$h = 1/4''$ flange angle thickness
$h/2 = 0.125''$
$c = 12.125''$
$f = 14''$
$d = 13.00''$
$e = 12.75''$ (gasket width = 0.75″)
$n = 16$ bolts
L_d = Length of duct, in. = 120 in. = distance that the vertical component of the bent duct reaction acts on flange.
$L_d/2$ = Distance that weight of duct acts on flange = 60 in.

Solving for the load from the moment induced by the bent duct vertical reaction, F_y: refer to Figure 6.20. The total bolt load, W_T, is equally distributed over the 16 bolts when the duct is straight and the weight of the duct is not included in the calculations. When the weight of the duct is included and the duct is bent, the duct bent force is directed at an angle and induces a moment in the flange, a shear load on the bolts and the bolt tension load is not equally distributed. Refer to the following examples.

Assume that the vent and duct is 24″ diameter. The flange is 2″ × 2″ × 1/4″ angle.

Assume that there is no column support for a conservative analysis. All of the load is taken by the flange and bolts.

The duct flange bolt arrangement is to match the explosion vent flange. For this example assume that there are 16(1/2)″ bolts (refer to Figure 6.20).

All duct flanges are matched to the explosion vent flange. The bolts attaching the first duct flange on the spool section to the vent flange are the most highly loaded and stressed bolts.

Combined shear and tension in the 1/2″ bolts: the maximum bolt loading is on the bolt, b9, farthest from point "A" at the top, distance d9. Summation of the moments about point "A" will provide the maximum load on bolt b9.

Where:

A_r = Stress area of 1/2″ bolt = 0.1597 in.^{2}
A_d = Full diameter shear area at 0.4387 minor diameter = 0.1512 in.^{2}
n = Number of bolts = 16.

$$M = \text{Venting load moment on bent duct} = F_y \times L_d = 2449 \times 120 = 293\,880 \text{ in.-lbs} \quad (6.136)$$

$$M_D = \text{Duct weight load moment} = F_d \times L_d/2 = 400 \times 60 = 24\,000 \text{ in.-lbs} \quad (6.137)$$

$$f' = 1/2\left(f_t + \sqrt{f_t^2 + 4f_s^2}\right) = \text{Combined bolt tension and shear stress, psi}$$

$$f_s = (F_y + F_d)/nA_d = (2449 + 400)/16 \times 0.1512 = 1178 \text{ psi bolt shear stress} \quad (6.138)$$

$f_t = F_b/A_r$ = Total axial bolt tension stress due to moment, psi.

The load for bolt b9 a distance d9 from point "A":

$$F_{wb} = \left[(F_y \times L_d) + F_d \times L_d/2\right]d9/(a + b)(na^2 + nb^2/2) \quad (6.139)$$

$$F_{wb} = [(2449 \times 120) + 400 \times 60]27/(16 \times 14^2 + 16 \times 13^2/2) = 317\,880(27)/4488 = 1917 \text{ lbs load on bolt b9} \quad (6.140)$$

$$w_b = 153 + 729 + 88 + 1917 = 2887 \text{ lbs/bolt} \quad (6.141)$$

$$F_t = W_b/A_r = 2887/0.1597 = 18\,077 \text{ psi/bolt} \quad (6.142)$$

Then,

$$f' = 1/2\left(18\,077 + \sqrt{18\,077^2 + 4 \times 1178^2}\right) = 1/2(18\,077 + 18\,229) = 18\,153 \text{ psi combined bolt tension and shear stress} \quad (6.143)$$

Refer to Table 5.1 for allowable bolt stress (Figure 6.21).

$$\text{The factor of safety} = 24\,120/18\,153 = 1.3 \quad (6.144)$$

$$\text{The required bolt torque is} = 0.2 \times 2887 \times 0.5 = 289 \text{ in.-lbs} \quad (6.145)$$

The worst bolt load on bolt b9 is 2887 lbs from the case where there is a 45° bend at the end of the duct.

$$\text{The span between bolts is} = \pi \times 26/16 \text{ bolts} = 5.1 \text{ in.} \quad (6.146)$$

Then,

$$F/\text{inch of flange} = 2887 \text{ lbs}/5.1 \text{ in.} = 566 \text{ lbs/in.} \quad (6.147)$$

$$I_{x-x} = bt_f^3/12 = 2(0.25^3)/12 = 0.0026 \text{ in.}^4 \quad (6.148)$$

$$M_t = F/\text{inch} \times b = 566 \times 2 = 1132 \text{ in.-lbs/in.} \quad (6.149)$$

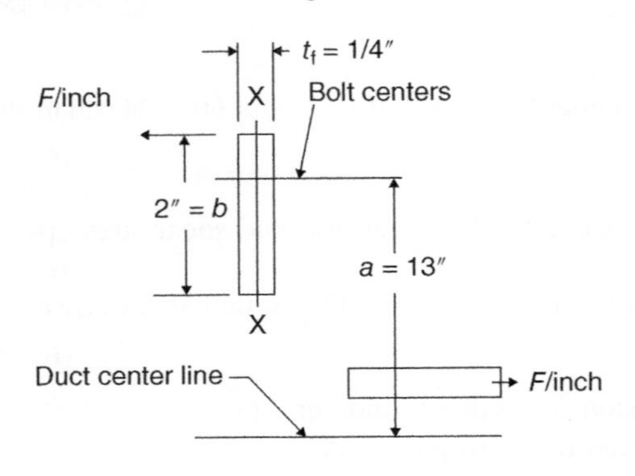

Figure 6.21 Round flange stress with a 45° bend at the end of the duct (for the round flange, refer to Figures 6.18, and 6.20).

$$C = t_f/2 = 0.25/2 = 0.125 \text{ in.} \tag{6.150}$$

$$\text{Stress} = M_t C/I_{x-x} = 1132(0.125)/0.0026 = 54\,423 \text{ psi} \tag{6.151}$$

$$\text{Factor of safety} = 24\,120/54\,423 = 0.44$$

The flange is overstressed and is not safe. Increase the thickness of the angle flange to 3/8″ (Figure 6.22).

The worst bolt load on bolt b9 is 2887 lbs from the case where there is a 45° bend at the end of the duct.

$$\text{The span between bolts is} = \pi \times 26/16 \text{ bolts} = 5.1 \text{ in} \tag{6.152}$$

Then,

$$F/\text{inch of flange} = 2887 \text{ lbs}/5.1 \text{ in.} = 566 \text{ lbs/in.} \tag{6.153}$$

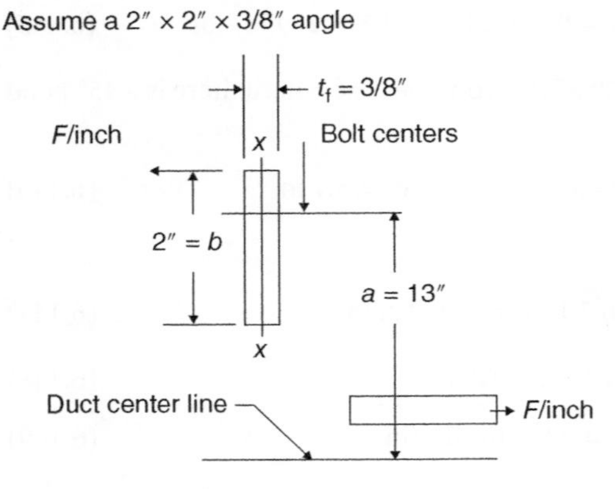

Figure 6.22 Flange stress with a 45° bend (for the round flange Refer to Figures 6.18, and 6.20).

$$I_{x-x} = bt_f^3/12 = 2(0.375^3)/12 = 0.0088\ \text{in.}^4 \quad (6.154)$$

$$M_t = F/\text{inch} \times b = 566 \times 2 = 1132\ \text{in.-lbs/in.} \quad (6.155)$$

$$C = t_f/2 = 0.375/2 = 0.188\ \text{in.} \quad (6.156)$$

$$\text{Stress} = M_t C/I_{x-x} = 1132(0.188)/0.0088 = 24\,183\ \text{psi} \quad (6.157)$$

$$\text{Factor of safety} = 24\,120/24\,183 = 0.997 \quad (6.158)$$

The flange is safe due to the conservative analysis where the allowable stress is 67% of the 0.2% yield strength of the material.

For a circular bolted flange with a 90° bend at the end of the duct and a 2″ × 2″ × 1/4″ angle flange (refer to Figures 6.19 and 6.20):

All of the conditions are the same except that $F_r = 0$ and F_{bd} is acting at the end of duct

$$W_T = \text{Maximum total flange load} = F_r + P' + F_g + F', \text{lbs} \quad (6.159)$$

$$W_b = \text{Maximum individual bolt load} = F_{rb} + P_b' + F_{gb} + F_{wb}, \text{lbs} \quad (6.160)$$

F_r = Venting reaction force = $a_u(\text{DLF})(A_d)P_{Red}$ (refer to Chapter 8 for reaction force calculation) = 3463 lbs. *Note*: This horizontal force is for a straight vent duct with no bend on the end.

Assume a 90° angle at the end of the duct with a 30 in. bend extension.

F_x = Horizontal component of the reaction force with a 90° bend at the end of the duct, lbs = 0 lbs.

F_y = Vertical component of the reaction force with a 90° bend at the end of the duct, lbs = 3463 lbs.

F_d = Duct weight = (π × duct diameter) × t_d × ($L_d + L_{dext}$) × 0.3 lbs/in.³, lbs

L_d = Length of horizontal duct = 120″

L_{dext} = Length of extended bent duct = 30″

$L_d/2$ = Length of duct divided in half, in.

$$F_d = \pi \times 24 \times 0.1345(120 + 30) \times 0.30 = 456\ \text{lbs} \quad (6.161)$$

F_d = Assume duct weight = 550 lbs with reinforcing

t_d = Duct wall thickness = 0.1345 in.

$$P' = \text{Total force to seal gasket} = \pi(e^2 - b^2) \times 200\ \text{psi} = \pi(12.75^2 - 12^2) \times 200 = 11\,663\ \text{lbs} \quad (6.162)$$

$$P_b' = \text{Bolt force required to seal gasket} = P'/n = 11\,663/16 = 729\ \text{lbs/bolt} \quad (6.163)$$

$$F_g = \text{Total load required to compress gasket } 10\% = 0.10(F_x + P') = 0.10(0 + 11\,663) = 1166\ \text{lbs} \quad (6.164)$$

$$F_{gb} = \text{Bolt force required to compress gasket} = F_g/n = 1166/16 = 73\ \text{lbs/bolt}$$

F_{wb} = Maximum bolt load resultant of the moment from the weight of duct, lbs.
F_{bd} = Maximum bolt load resultant of the moment from the bent duct, lbs.
$b = 12''$ (for a 24″ duct)
$h = 1/4''$ flange angle thickness
$h/2 = 0.125''$
$c = 12.125''$
$f = 14''$
$d = 13.00''$
$e = 12.75''$ (gasket width = 0.75″)
$n = 16$ bolts

L_d = Length of duct = 120 in. = distance that the vertical component of the bent 90°duct reaction acts on the flange

$L_d/2$ = Distance that weight of duct acts on the flange = 60 in.

Solving for the load from the moment induced by the 90° bent duct vertical reaction, F_y: refer to Figure 6.20. The total bolt load, W_T, is equally distributed over the 16 bolts when the duct is straight and the weight of the duct is not included in the calculations. When the weight of the duct is included and the duct is bent the duct bent force is directed at an angle and induces a moment in the flange, a shear load on the bolts and the bolt load is not equally distributed. Refer to the following examples.

Assume that the vent and duct is 24″ diameter. The flange is 2″ × 2″ × 1/4″ angle.

Assume that there is no column support for a conservative analysis. All of the load is taken by the flange and bolts.

The duct flange bolt arrangement is to match the explosion vent flange. For this example assume that there are 16(1/2)″ bolts (refer to Figure 6.20).

All duct flanges are matched to the explosion vent flange. The bolts attaching the first duct flange on the spool section to the vent flange are the most highly loaded and stressed bolts.Combined shear and tension in the 1/2″ bolts: the maximum bolt loading is on the bolt, b9, farthest from point “A” at the top, distance d9. Summation of the moments about point “A” will provide the maximum load on bolt b9.

Where

A_r = Stress area of 1/2″ bolt = 0.1597 in.^{2}
A_d = Full diameter shear area at 0.4387 minor diameter = 0.1512 in.^{2}
n = Number of bolts = 16.

$$\begin{aligned} \mathrm{M} &= \text{Venting load moment on 90°bent duct} = F_y \times L_d \\ &= 3463 \times 120 = 415\,560 \text{ in.-lbs} \end{aligned} \tag{6.165}$$

$$\begin{aligned} M_d &= \text{Duct weight load moment} = F_d \times L_d/2 \\ &= 550 \times 60 = 33\,000 \text{ in.-lbs} \end{aligned} \tag{6.166}$$

$$f' = 1/2\left(f_t + \sqrt{f_t^{\,2} + 4f_s^{\,2}} \right) = \text{combined bolt tension and shear stress, psi} \tag{6.167}$$

$$f_s = F_y + F_d/nA_r = \text{Bolt shear stress} = (3463 + 550)/(16 \times 0.1512) = 1658 \text{ psi} \tag{6.168}$$

$$f_t = F_b/A_r = \text{Total axial bolt tension stress due to moment, psi} \tag{6.169}$$

The load for bolt b9 a distance d9 from point "A":

$$F_{wb} = \left(F_y \times L_d + F_d \times L_d/2\right)\mathrm{d}9/\left(a + b/na^2 + nb^2/2\right) = \text{Load on bolt b9} \tag{6.170}$$

$$\begin{aligned} F_{wb} &= (3463 \times 120 + 550 \times 60)27/(16 + 14)/16 \times 14^2 + 16 \times 13^2/2 \\ &= (448\,560)27/4488 = 2698 \text{ lbs load on bolt b9} \end{aligned} \tag{6.171}$$

$$W_b = 0 + 729 + 73 + 2698 = 3500 \text{ lbs on bolt b9} \tag{6.172}$$

$$F_t = W_b/A_r = 3500/0.1597 = 21\,916 \text{ psi} \tag{6.173}$$

Then,

$$\begin{aligned} f' &= 1/2\left(21\,916 + \sqrt{21\,916^2 + 4 \times 1597^2}\right) = 1/2(21\,916 + 22\,147) \\ &= 22\,031 \text{ psi combined bolt tension and shear stress} \end{aligned} \tag{6.174}$$

Refer to Table 5.1 for allowable bolt stress (Figure 6.23).

$$\text{The factor of safety} = 24\,120/22\,031 = 1.09 \tag{6.175}$$

$$\text{The required bolt torque is} = 0.2 \times 3500 \times 0.5 = 350 \text{ in.-lbs} \tag{6.176}$$

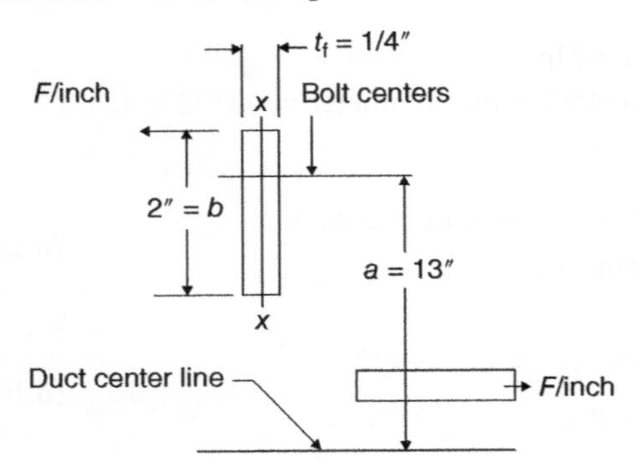

Figure 6.23 Flange stress for the round flange with 90° bend at end of duct (refer to Figures 6.19 and 6.20).

The worst bolt load on bolt b9 is 3500 lbs from the case where there is a 90° bend at the end of the duct.

The span between bolts is $\pi \times 26''/16$ bolts $= 5.1''$

Then,

$$F/\text{inch} = \text{The load on the flange} = 3500\ \text{lbs}/5.1 = 686\ \text{lbs/in.} \tag{6.177}$$

$$I_{x-x} = bt_f{}^3/12 = 2(0.25^3)/12 = 0.0026\ \text{in.}^4 \tag{6.178}$$

$$M_t = F/\text{inch} \times b = 686 \times 2 = 1372\ \text{in.-lbs/in.} \tag{6.179}$$

$C = t_f/2 = 0.25/2 = 0.125$ in.

$$\text{Stress} = M_t C/I_{x-x} = 1372(0.125)/0.0026 = 65\,961\ \text{psi} \tag{6.180}$$

$$\text{Factor of safety} = 24\,120/65\,961 = 0.36 \tag{6.181}$$

The flange is overstressed and is not safe. Increase the thickness of the angle flange to 3/8″ (Figure 6.24).

The worst bolt load on bolt b9 is 3500 lbs from the case where there is a 90° bend at the end of the duct.

The span between bolts is = 5.1″

Then, F/in. = the load on the flange is = 686 lbs/in.

$$I_{x-x} = bt^3/12 = 2(0.375^3)/12 = 0.0088\ \text{in.}^4 \tag{6.182}$$

$$M_t = F/\text{inch} \times b = 686 \times 2 = 1372\ \text{in.-lbs/in.} \tag{6.183}$$

$$C = t_d/2 = 0.375/2 = 0.188\ \text{in.} \tag{6.184}$$

$$\text{Stress} = M_t C/I_{x-x} = 1372(0.188)/0.0088 = 29\,310\ \text{psi} \tag{6.185}$$

$$\text{Factor of safety} = 24\,120/29\,310 = 0.82 \tag{6.186}$$

The flange is overstressed and not safe. Increase the angle flange thickness to 7/16″ (Figure 6.25).

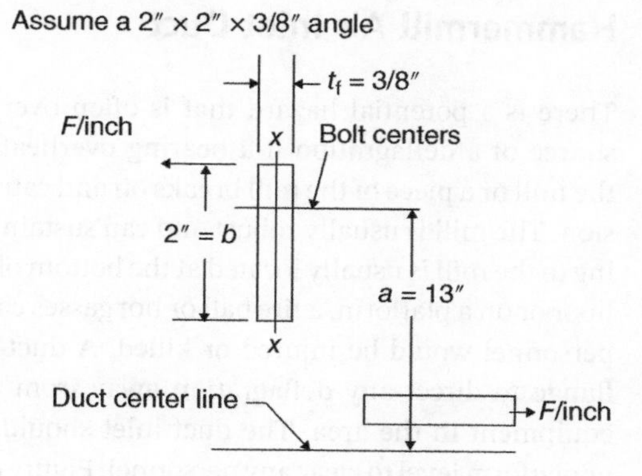

Figure 6.24 Flange stress (for the round flange refer to Figures 6.19 and 6.20).

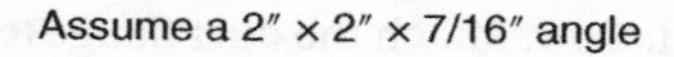

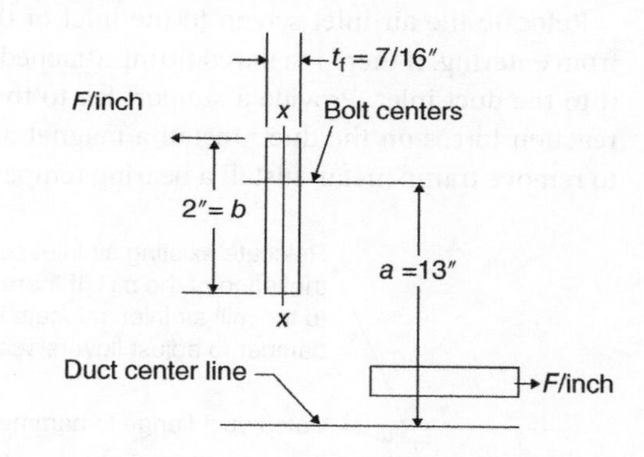

Figure 6.25 Flange stress (for the round flange refer to Figures 6.19 and 6.20).

The worst bolt load on bolt b9 is 3500 lbs from the case where there is a 90° bend at the end of the duct.

The span between bolts is = 5.1″

Then, F/in. = The load on the flange is = 686 lbs/in.

$$I_{x-x} = bt^3/12 = 2(0.438^3)/12 = 0.014 \text{ in.}^4 \tag{6.187}$$

$$M_t = F/\text{inch} \times b = 686 \times 2 = 1372 \text{ in.-lbs/in.} \tag{6.188}$$

$$C = t_d/2 = 0.438/2 = 0.219 \text{ in.} \tag{6.189}$$

$$\text{Stress} = M_t C/I_{x-x} = 1372(0.219)/0.014 = 21\,462 \text{ psi} \tag{6.190}$$

$$\text{Factor of safety} = 24\,120/21\,462 = 1.12 \tag{6.191}$$

The flange is safe.

Hammermill Air Inlet Duct

There is a potential hazard that is often overlooked. The hammermill can be a source of a deflagration if a bearing overheats or a piece of tramp metal enters the mill or a piece of the mill breaks off and causes friction that may cause an explosion. The mill is usually robust and can sustain high pressure. The air intake opening to the mill is usually located at the bottom of the mill. If the mill is located on the floor or on a platform, a fire ball or hot gasses can be discharged into the area where personnel would be injured or killed. A duct should be attached to the air inlet flange to direct any deflagration away from danger to personnel or vulnerable equipment in the area. The duct inlet should be located at least 7 ft above floor or platform level to clear any personnel. Figure 6.26 illustrates a recommended duct with a large radius sweep elbow to reduce any pressure drop from interfering with the required inlet air flow. Match the duct flange to the air inlet flange of the mill.

Relocate the air inlet screen to the inlet of the duct to prevent foreign objects from entering. If there is a flared fitting attached to the mill air inlet flange relocate it to the duct inlet. Provide a support leg to the duct as shown to withstand any reaction forces on the duct. Install a magnet above the product inlet to the mill to remove tramp metal. Install a bearing temperature sensor and vibration switch

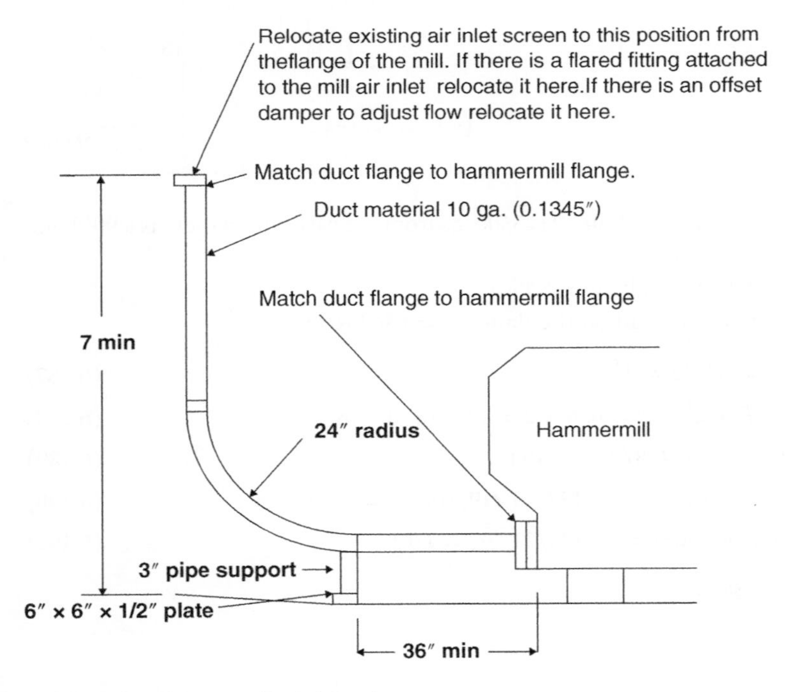

Figure 6.26 Hammermill air inlet duct.

on the mill. Alarm the bearing temperature sensor to shut down the mill when the normal steady-state reading increases by 30 °C. Calibrate the vibration switch to shut down the mill when the normal amplitude of vibration increases by 25% for more than one minute.

Provide an annular flow area to keep bags from blocking flow. Use 1/4″ steel rods spaced approximately 4″ apart. For the 24″ × 24″ vent example (Figure 6.27).

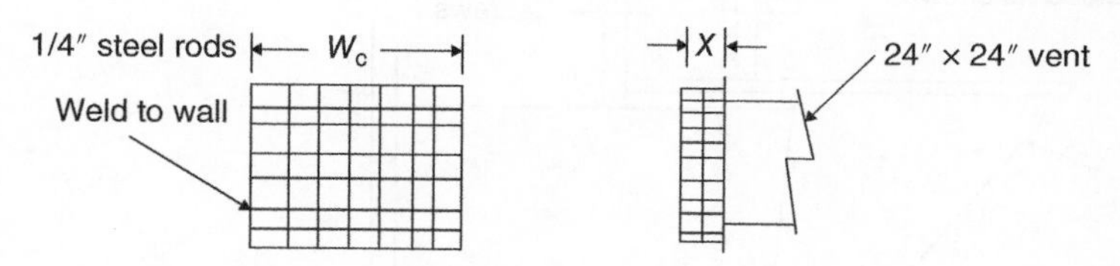

Figure 6.27 Filter bag cage detail.

The cage for a 24″ square vent is 28″ × 28″. The annular flow area, A_f, for a 24″ × 24″ vent = 576 in.^{2}

$A_f = (4 \times Wc)$, X = the annular perimeter area of the cage. $Wc = 28$, then $X = 576/4 \times 28 = 5.2''$ minimum. Make $X = 6.0''$.

If any filter bags interfere with the filter bag cage near the vent opening, remove the interfering bags in the area and blank off the holes left in the tube sheet with welded plate.

Figure 6.28 illustrates a wall penetration for a non-loss-in-weight vessel.

Figure 6.29 illustrates a wall penetration for a loss-in-weight vessel.

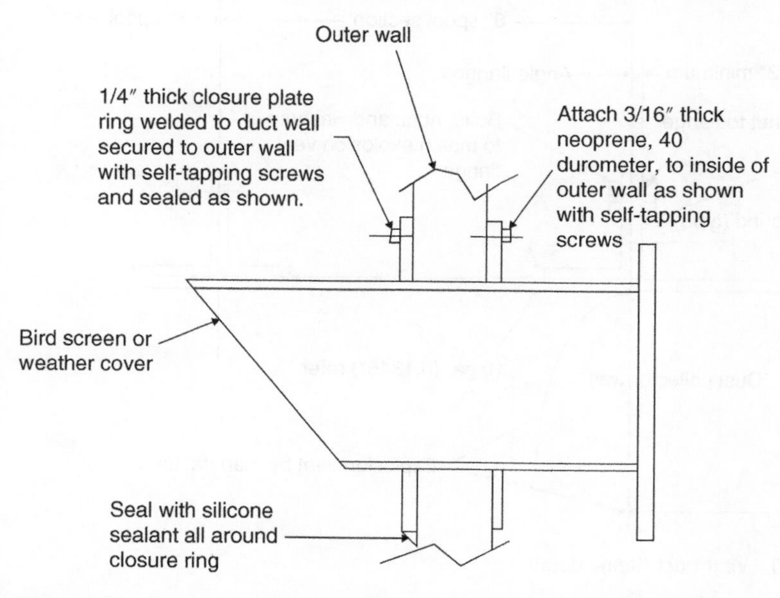

Figure 6.28 Wall penetration vent duct non-loss-in-weight vessel.

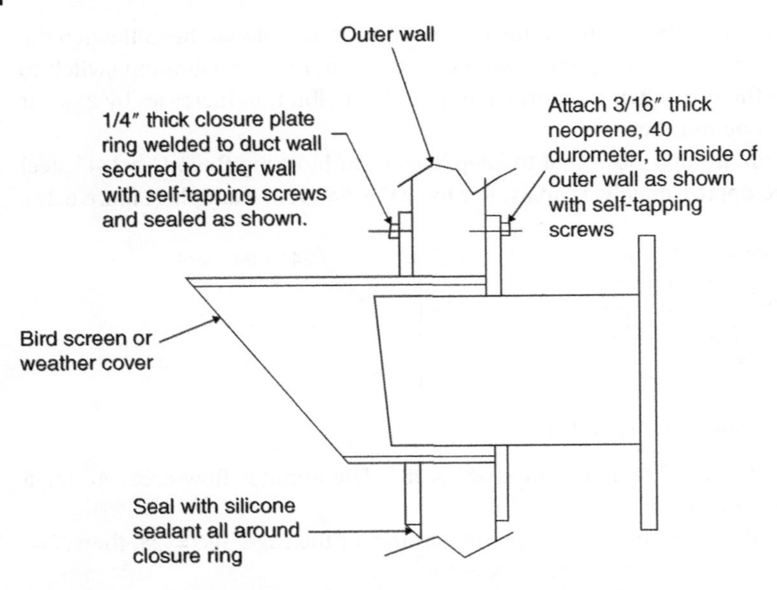

Figure 6.29 Wall penetration vent duct loss-in-weight vessel.

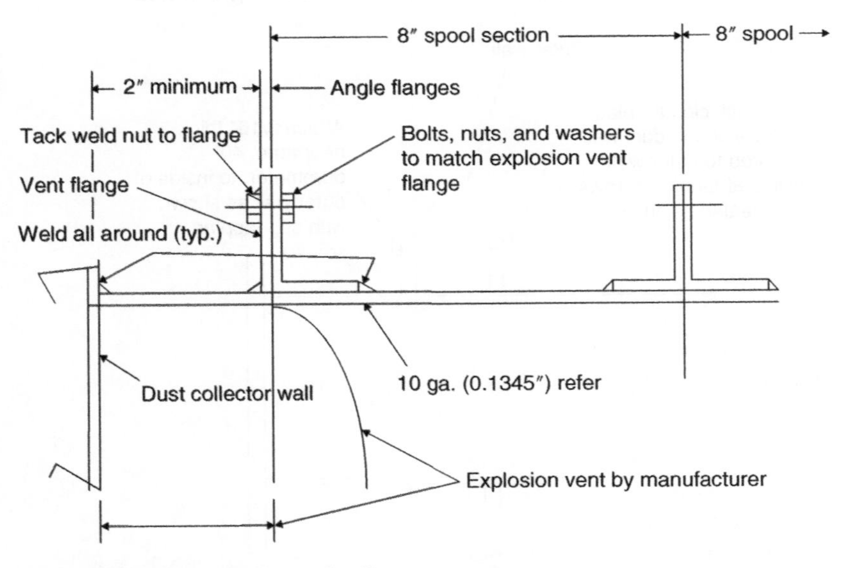

Figure 6.30 Vent duct flange detail.

1/4″ thick closure plate welded to the outer duct and secured to wall with self-tapping screws and sealed with silicone sealant as shown.

Mount 3/16″ thick, durometer 40 neoprene, to inside of wall with self-tapping screws. The inside dimension of seal to be 1/4″ less than the outside dimension of vent duct. The construction of the outside wall varies with the type of building; therefore, the connection to the wall may vary (Figure 6.30).

Bird Screens

Bird screens are to be provided on blast deflectors to keep animals and birds from entering the duct. Figure 6.31 illustrates the bird screen design.

It is important that the effective venting flow area of the duct is not unduly impeded by the bird screen. The size of the wires and the spacing has been designed to ensure that flow area with the screen is greater than the area of the duct to ensure that flow restriction is kept to a minimum.

For a 24″ × 24″ duct, the flow area is 576 in.^{2}. The screen area must then be larger to account for the screen open area being 83.9%. The blocked area caused by the screen is 100 – 83.9 = 16.1%. The required screen flow area is 24″ × L_s = 576 × 1.161 = 668 in.^{2}. L_s = 668/24 = 27.8″ minimum. Make L_s = 30″.

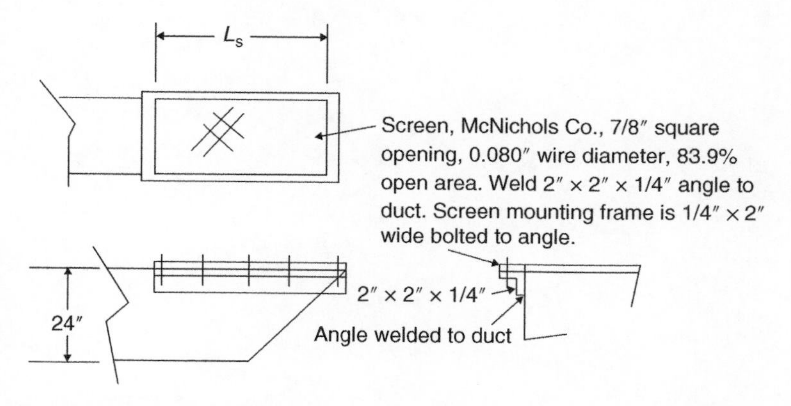

Figure 6.31 Bird screen detail.

7

Explosion Vent Duct Weather Covers

Chapter 6 has provided the proper design for the explosion vent duct. The explosion vent duct has been directed through a wall to the outside or through the roof. Consideration has been given to ensure that the gasses being vented are directed away from any personnel access or vulnerable equipment nearby.

The weather covers are to be used wherever rain, snow, or animals can enter the duct. Experience has shown that the most cost-effective and reliable design is a polystyrene weather cover. The proprietary weather covers are illustrated on Figures 7.1 and 7.4 for weather cover frame and support rods, Figure 7.2 for weather cover on square cut round ducts, Figure 7.3 for bevel cut round ducts, Figure 7.6 for weather cover on square cut rectangular ducts, and Figure 7.5 for weather cover on bevel cut rectangular ducts.

These weather covers have been designed to snow, ice, and wind force design criteria. With a horizontal weather cover, the depth of snow must be limited to 3.75 in. and the depth of ice must be limited to 0.53 in. to keep the weight of the snow or ice below 2.5 lbs/ft^2 (the acceptable limit of restraint on the weather cover to allow unrestricted venting). A flat horizontal weather cover would not shed snow or ice; therefore, the weather cover must be placed on a duct bevel cut at a 45° angle to ensure that accumulation of snow or ice is kept to a minimum. External flat horizontal duct weather covers are only to be used where snow or ice accumulations are not possible.

The maximum wind force is assumed to be 100 mph (the worst case wind gust country wide). With the wind on the windward side of the duct, there is a suction pressure on the lee side that must be resisted by the weather cover disk. The suction pressure is 0.18 psi for these conditions. The venting pressure that fractures the weather cover is 0.5 psi; therefore, if the disk is designed to withstand 0.18 psi under normal conditions, the higher venting pressure will fracture the disk easily.

Figures 7.2, 7.3, 7.5, and 7.6 provide the disk thickness required for each duct size and shape.

Explosion Vented Equipment System Protection Guide, First Edition. Robert C. Comer.

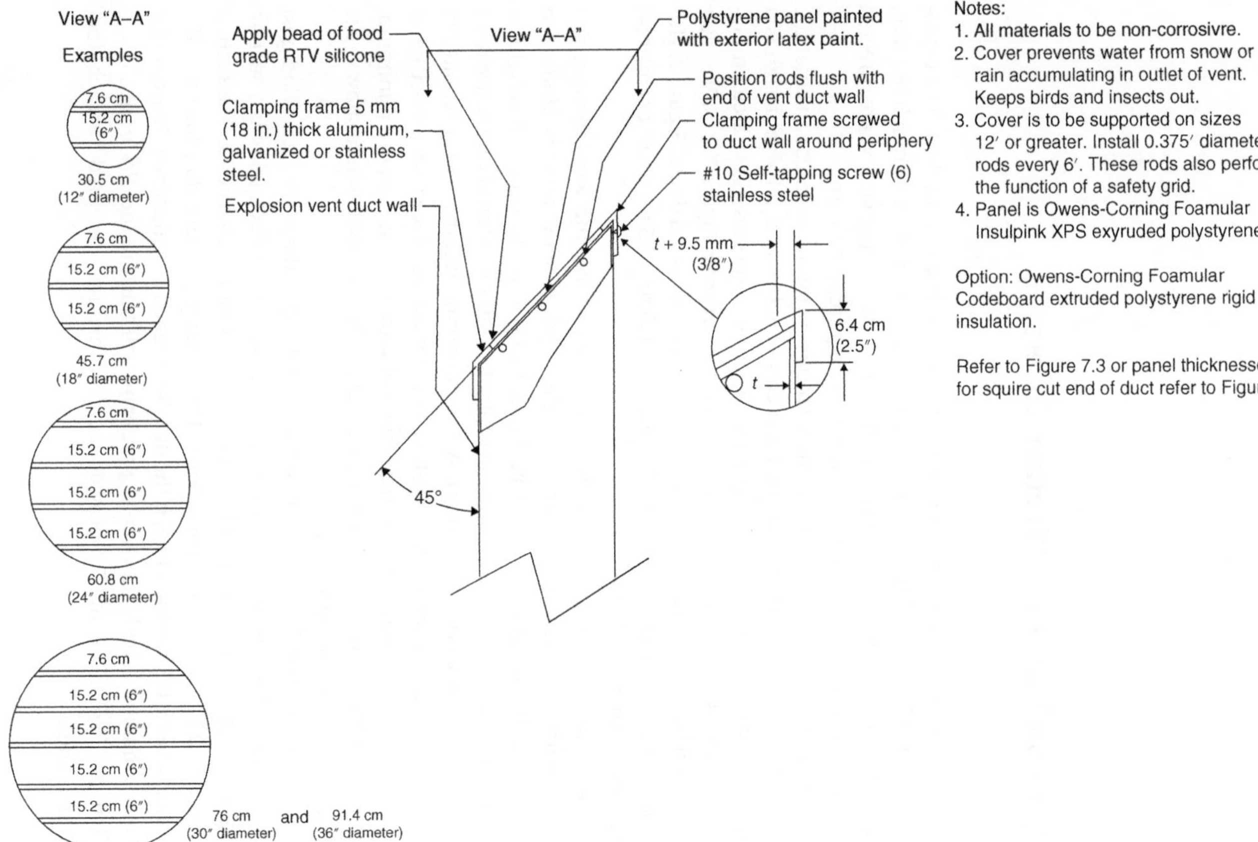

Figure 7.1 Support frame bevel cut weather cover round duct.

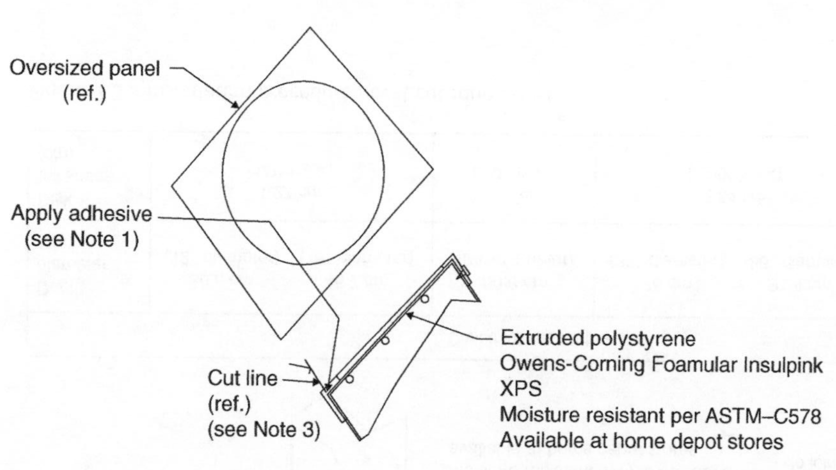

Installation procedure:

1. Apply solvent free epoxy adhesive (Lepage's metal epoxy or J.B Weld) to duct wall edge.
2. Place oversized extruded polystyrene panel on duct wall and allow to dry and retain panel.
3. Use duct wall as a guide and cut polystyrene panel with a saw or hot knife (Uline co., foam hot knife, 110 V, model H–1079. 10 cm (4″) blade).
4. Paint exposed surfaces of polystyrene disc with an exterior latex paint
5. Place aluminum clamping frame over disc and screw to duct wall with (6) #10 self-tapping stainless steel screws.

Disc thickness table						
Duct diameter (cm)	30.5 cm (12″ diameter)	45.7 cm (18″ diameter)	60.8 cm (24″ diameter)	76 cm (30″ diameter)	91.4 cm (36″ diameter)	101.6 cm (40″ diameter)
Disc thickness (cm)	0.63 cm (1/4″)	1.27 cm (1/2″)	1.27 cm (1/2″)	1.90 cm (3/4″)	2.22 cm (7/8″)	2.54 cm (1.00″)

Figure 7.2 Installation procedure square cut weather cover round duct.

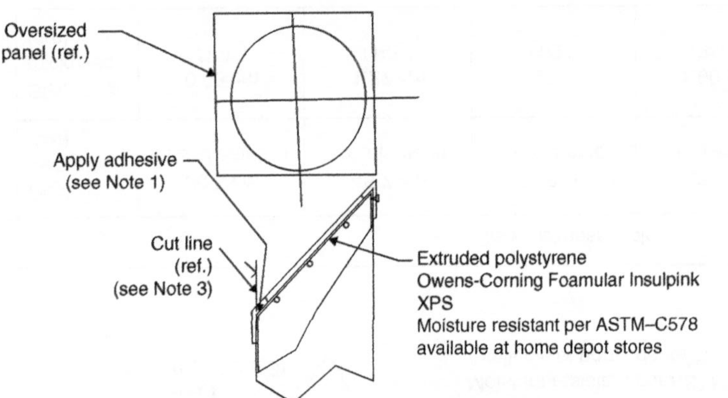

Installation procedure:

1. Apply solvent free epoxy adhesive LePage's Metal Epoxy or J.B.Weld to duct wall edge.
2. Place oversized extruded polystyrene panel on duct wall and allow to dry and retain panel.
3. Use duct wall as a guide and cut polystyrene panel with a saw or hot knife Uline co., Foam Hot Knife, 110 v, Model H–1079. 10 cm 4″, blade.
4. Paint exposed surfaces of polystyrene disk with an exterior latex paint
5. Place clamping frame over disk, press down firmly to provide a mild clamping force to the disk and screw to duct. wall with 6 number 10 self-tapping stainless steel screws.
6. Apply bead of food grade RTV silicone sealant to joint between frame and painted disk.

Disk thickness table				
Duct diameter (cm)	30.5 cm (12″ diameter) 45.7 cm (18″ diameter)	60.8 cm (24″ diameter)	76 cm (30″ diameter) 91.4 cm (36″ diameter)	101.6 cm (40″ diameter)
Disk thickness (cm)	1.27 cm (1/2″ thick)	1.9 cm (3/4″ thick)	2.54 cm (1.00″ thick)	3.81 cm (1–1/2″ thick)

Figure 7.3 Installation procedure bevel cut round duct.

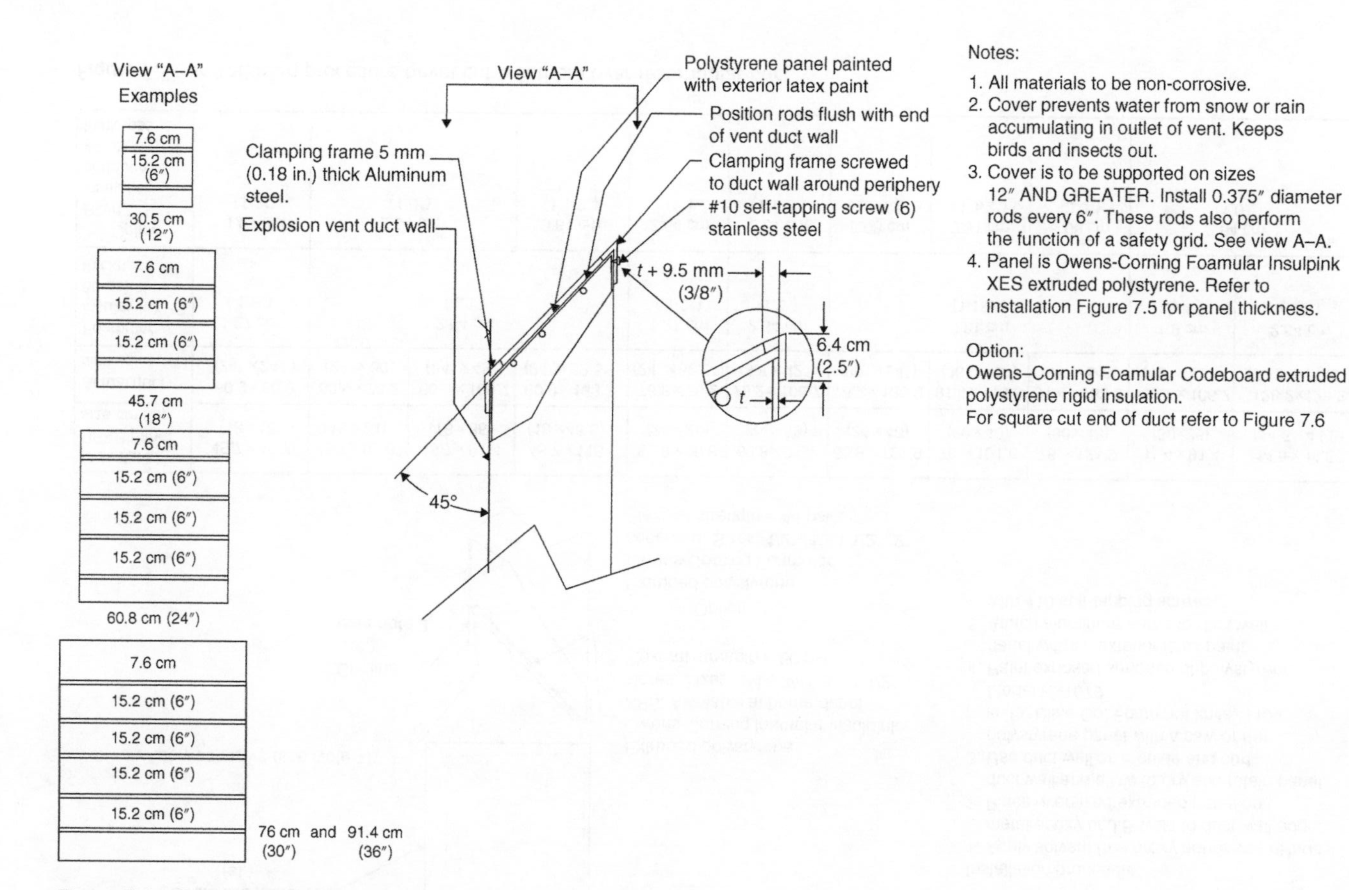

Figure 7.4 Support frame bevel cut weather cover rectangular duct.

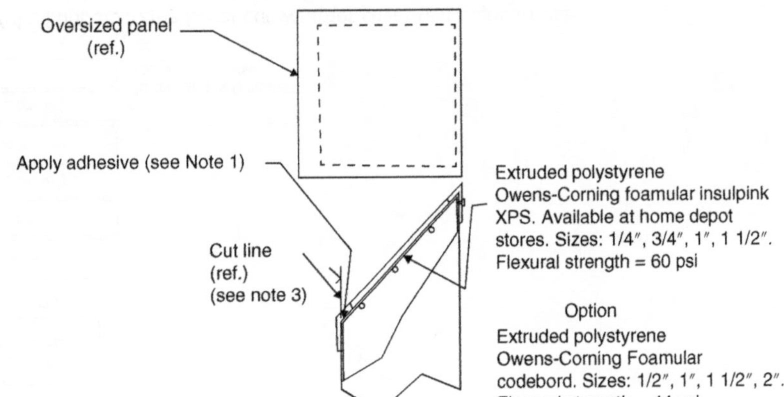

Installation procedure:
1. Apply solvent free eroxy adhesive LePage's metal epoxy or J.B weld to duct wall edge.
2. Place oversized extruded panel on duct wall and allow to dry and retain panel.
3. Use duct wall as a guide and curt polystyrene panel with a saw or hot knife, Uline Co. Foam hot knife, 110v, Model H–1079.
4. Paint exposed surfeace of polystyrene panel with an exterior latex paint.
5. Attach aluminum frame to duct wall with #10 self-tapping screws.

Calculated BS&B panel size cm (in)	45.7 × 45.7 (18 × 18)	45.7 × 60.8 (18 × 24)	45.7 × 91.4 (18 × 36)	45.7 × 118 (18 × 46.5)	60.8 × 60.8 (24 × 24)	60.8 × 91.4 (24 × 36)	60.8 × 121.9 (24 × 48)	76. × 101.6 (30 × 40)	76. × 121.9 (30 × 48)	91.4 × 91.4 (30×36)	(44.5 × 44.5″) (44.5 × 44.5)
Actual duct size	60.9 × 60.9 (24″ × 24″)	60.9 × 76.2 (24″ × 30)	60.9 × 106.7 (24″ × 42″)	60.9 × 133.3 (24″ × 52.5″)	76.2 × 76.2 (24″ × 52.5″)	76.2 × 106.7 (30″ × 42″)	76.2 × 133.3 (30″ × 54″)	91.4 × 116.8 (36″ × 46″)	91.4 × 133.3 (36″ × 54″)	106.7 × 106.7 (42″ × 42″)	128.2×128.2 (50.5″ × 50.5″)
Foamular insulpink XPS 60 psi thickness	1.27 cm (1/2″)	2.54 cm (1.0″)			1.27 cm (1/2″)	2.54 cm (1.0″)	3.81 cm (1 1/2″)			1.9 cm (3/4″)	2.54 cm (1.0″)
Option Rigid insulation codebord 44 psi thickness	1.27 cm (1/2″)	2.54 cm (1.0″)		3.81 cm (1 1/2″)	2.54 cm (1″)	3.81 cm (1 1/2″)	5.08 cm (2.0″)	3.81 cm (1 1/2″)	5.08 cm (2.0″)	2.54 cm (1.0″)	

Figure 7.5 Installation procedure bevel cut weather cover rectangular duct.

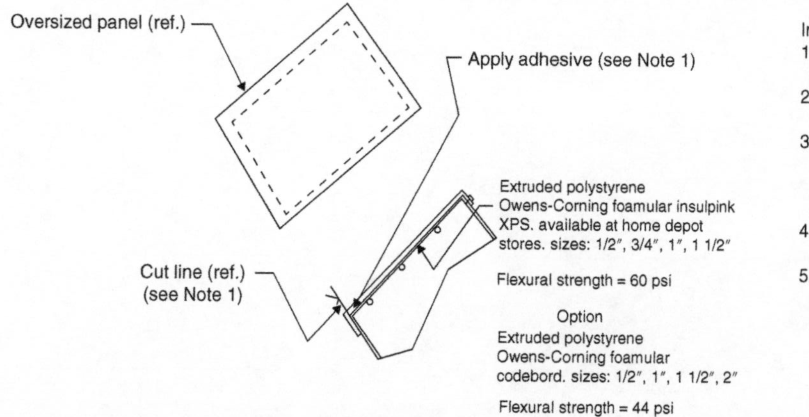

Installation procedure:

1. Apply solvent free epoxy adhesive (Lepage's metal epoxy or J.B. Weld) to duct wall edge.
2. Place oversized extruded polystyrene panel on duct wall and allow to dry and retain panel
3. Use duct wall as a guide and cut polystyrene panel with a saw or hot knife (Uline co., foam hot knife, 110 V, model H–1079. 10 cm (4″) blade).
4. Paint exposed surfaces of polystyrene panel with an exterior latex paint
5. Place aluminum clamping frame over disc and screw to duct wall with (6) #10 self-tapping stainless steel screws.

Calculated BS&B panel size cm (in)	45.7×45.7 (18×18)	45.7×60.8 (18×24)	45.7×91.4 (18×36)	45.7×118 (18×46.5)	60.8×60.8 (24×24)	60.8×91.4 (24×36)	60.8×121.9 (24×48)	76.×101.6 (30×40)	76.×121.9 (30×48)	91.4×91.4 (36×36)	(44.5×44.5″) (44.5×44.5)
Actual duct size	60.9×60.9 (24″×24″)	60.9×76.2 (24″×30)	60.9×106.7 (24″×42″)	60.9×133.3 (24″×52.5″)	76.2×76.2 (30″×30″)	76.2×106.7 (30″×42″)	76.2×133.3 (30″×54″)	91.4×116.8 (36″×46″)	91.4×133.3 (36″×54″)	106.7×106.7 (42″×42″)	128.2×128.2 (50.5″×50.5″)
Foamular Insulpink XPS 60 psi thickness	1.27 cm (1/2″)						1.90 cm (3/4″)				
Option Rigid insulation codebord 44 psi thickness	1.27 cm (1/2″)			2.54 cm (1″)	1.27 cm (1/2″)	2.54 cm (1.0″)					

Figure 7.6 Installation procedure square cut weather cover rectangular duct.

The weather cover disk material is Owens-Corning Foamular Insulpink XPS Extruded Polystyrene. This material is available at Home Depot stores in thicknesses of 1/2″, 3/4″, 1.00″, and 1-1/2″ in 4 ft × 8 ft sheets. The disk can be cut to fit the shape of the duct wall with a saw or a hot knife using the duct wall edge as a guide after using a solvent free epoxy adhesive to retain the panel to the duct wall. Refer to Figures 7.1 and 7.4 for the installation procedure.

The surfaces of the finished disk exposed to the weather are to be painted with an exterior latex paint.

8

Dust Collector Stability

The dust collector supporting structure must be strong enough and stabile to withstand any reaction forces that develop as a result of operation of the explosion vent, including the dynamic effect of the rate of force application as expressed by a dynamic load factor (DLF). The total reaction force shall be the vector sum of each force component applied at the geometric center of each vent.

The reaction force at each vent opening **without** a duct is given by the following:

$$F_r = \text{au}(\text{DLF})(A_v)P_{\text{Red}} \tag{8.1}$$

where au = units conversion: 100 for SI units; 1 for customary units.

$$\text{DLF} = 1.2 = \text{Dynamic load factor}$$

$$\begin{aligned} A_v &= \text{Vent area, m}^2 = 24 \times 24 = 576\ \text{in.}^2/144\ \text{in.}^2/\text{ft}^2 \\ &= 4\ \text{ft}^2 \times 0.0929\ \text{m}^2/\text{ft}^2 = 0.372\ \text{m}^2 \end{aligned} \tag{8.2}$$

$$P_{\text{Red}} = \text{Reduced pressure, bar}$$

$$F_r = 100(1.2)(0.372) \times 5.0\ \text{psi}/14.5\ \text{psi/bar} = 15.39\ \text{KN} \times 225\ \text{lbs/KN} = 3463\ \text{lbs} \tag{8.3}$$

The reaction force at the vent opening **with** a duct is given by the following:

Note: The reaction force **with** a duct is the same as the reaction force without a duct if the vent area is the same as the duct area. If the duct is larger than the vent, use the duct area. If the duct is bent at the end, the horizontal reaction, F_h, is used as defined in prior calculations and Eq. (8.5).

For dust collectors located on the inside of a building or enclosure, the reaction force due to the vented explosion is the only force to be considered; however, if the dust collector is located on the outside of a building, the potential wind overturning load must be factored in (Figure 8.1).

Explosion Vented Equipment System Protection Guide, First Edition. Robert C. Comer.

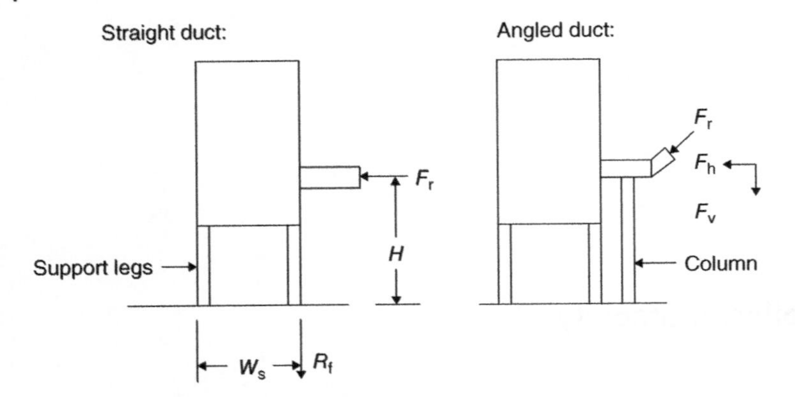

Figure 8.1 Inside location of dust collector: with vent duct.

With a straight duct, the overturning moment,

$$M_o = F_r H = \text{in.-lbs} \tag{8.4}$$

With an angled duct at 45°

$$F_h = F_v = F_r/\sqrt{2} \tag{8.5}$$

and the overturning moment,

$$M_o = F_h H = \text{in.-lbs} \tag{8.6}$$

The overturning moment is resisted by the support legs attached to the floor.

$$\text{Resisting force, } R_f = F_r H/W_s \tag{8.7}$$

The vertical component of the force, F_v, acting on the duct may be supported by a column to the floor or ground or by a strut to the dust collector framework. The dust collector may not be capable of sustaining the vent duct load on the duct connection flanges.

For the example: Refer to Figure 2.1

The distance from the floor to the center of the explosion vent is approximately 90 in. And $F_r = 3463$ lbs (worst case with a straight duct).

The overturning moment is then:

$$3463 \times 90 = 311\ 670 \text{ in.-lbs} \tag{8.8}$$

$$\text{The resisting force, } R_f = 311\ 670/W_s = 311\,670/33.125'' = 9408 \text{ lbs} \tag{8.9}$$

due to the explosion relief vent duct (Figure 8.2).

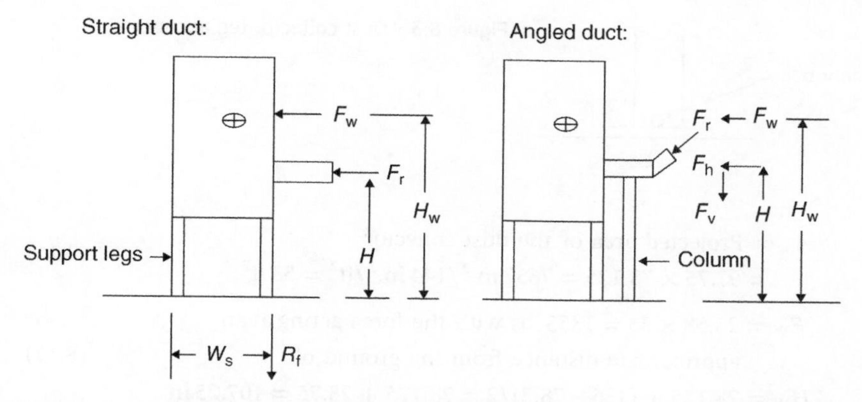

Figure 8.2 Outside location of dust collector: with vent duct.

With a straight duct, the overturning moment,

$$M_o = F_r H + F_W H_W = \text{in.-lbs} \quad (8.10)$$

with an angled duct normally at 45°

$$F_h = F_v = F_r/\sqrt{2} \quad (8.11)$$

and the overturning moment,

$$M_o = F_h H + F_W H_W = \text{in.-lbs} \quad (8.12)$$

The overturning moment is resisted by the support legs attached to the floor or ground.

$$\text{Resisting force, } R_f = (F_r H + F_W H_W)/W_s \quad (8.13)$$

The vertical component of the force, F_v, must be supported by a column to the floor or ground or by a strut to the dust collector framework.

Wind Force (F_w)

Assume 100 mi/h wind velocity at standard density (worst case) (Figure 8.3).

$$\text{Stagnation pressure for 100 mph wind} = q = 25.58\ \text{lbs/ft}^2 \quad (8.14)$$

(Refer Baumeister 1958)

$$F_W = q(A_{dc}) \text{ where } A_{dc} = \text{Area of dust collector exposed to the wind}) = \text{lbs} \quad (8.15)$$

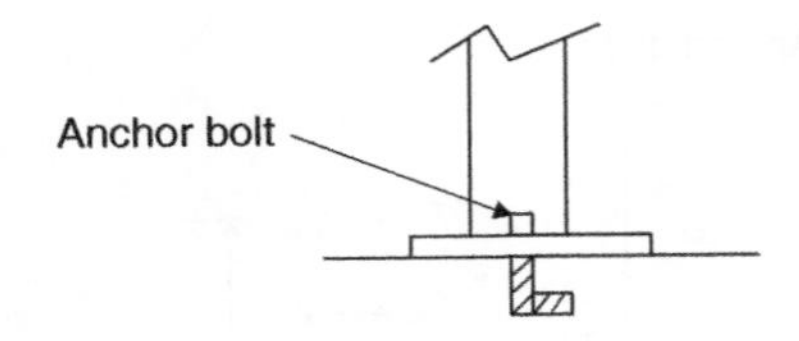

Figure 8.3 Dust collector leg anchors.

$$A_{dc} = \text{Projected area of the dust collector} = 97.75 \times 78.125 = 7637\ \text{in.}^2/144\ \text{in.}^2/\text{ft}^2 = 53\ \text{ft}^2 \quad (8.16)$$

$$F_W = 25.58 \times 53 = 1355\ \text{lbs with the force acting at an approximate distance from the ground of} \quad (8.17)$$

$$H_W = 78.125 + (136 - 78.5)/2 = 78.125 + 28.75 = 107.25\ \text{in.}$$

The overturning moment due to the wind is then 1325 × 107.125 = 145 154 in.-lbs The resisting force, R_f, due to the overturning moment of wind force is then:

$$R_f = 145\ \ 154/33.125 = 4382\ \text{lbs} \quad (8.18)$$

The total resisting force due to the explosion vent duct and the wind force acting on the two legs is then 9408 + 4382 = 137 90 lbs (8.19)

Each of the two legs must resist the uplifting force of 13790/2 = 6895 lbs (8.20)

Floor or Ground Anchor

The base plate in the example is 7-1/2″ × 7-1/2″ × 1/2″ thick with a 1″ diameter hole in the center. A 3/4″ diameter anchor bolt is used. Other base plates may have four 1/2″ diameter holes for 3/8″ diameter anchor bolts.

Assume that a 3/4″ diameter bolt is anchoring the dust collector to the roof, floor, or ground. The allowable tensile load for the weakest bolt material is based on a 67 % of the 0.2 % yield strength = 24 140 psi (Figure 8.4). The applied stress is 6895 lbs/0.4418 in.^{2} = 15 607 psi· The bolt area based on the nominal diameter of 3/4″ = 15 607 psi.

The factor of safety is then 24 140/15 607 = 1.55 (8.21)

If the anchor bolts were four 3/8″ diameter bolts, the applied stress per bolt would be

$$6895/4 \times 0.294 = 5863\ \text{psi} \quad (8.22)$$

with a factor of safety of

$$24\,140/58\,632 = 4.1 \quad (8.23)$$

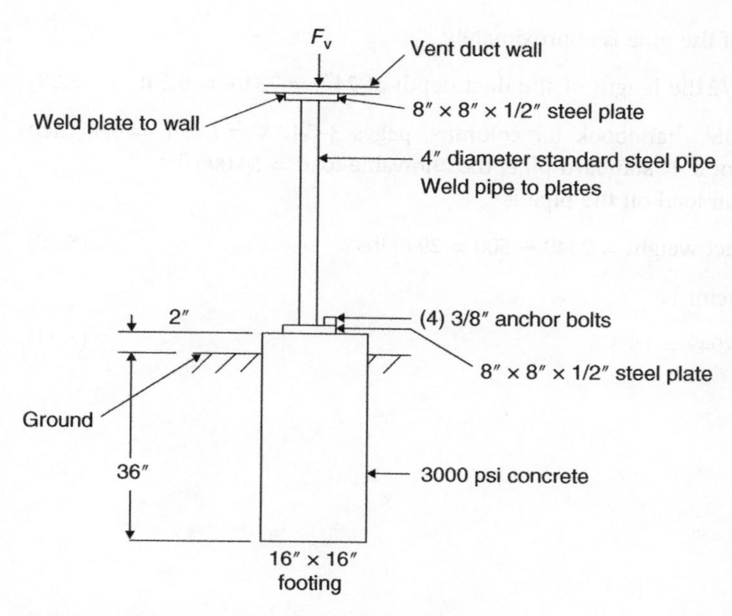

Figure 8.4 Angled vent duct column support.

The downward venting force acting on the column is

$$F_v = F_r/\sqrt{2} = 3463/1.414 = 2449 \text{ lbs} \quad (8.24)$$

Allowable approximate soil load = 4000 lbs/ft² min.
(Refer Baumeister 1958).

The footing concrete weight $= (36 \times 16 \times 16)/1728 \text{ in.}^3/\text{ft}^3 \times 193 \text{ lbs/ft}^3 = 1029 \text{ lbs}$ (8.25)

The vent duct approximate weight is

$$4 \times 24''/\text{side} \times 10 \text{ ga.}(0.1345) \times 120 \text{ in. long} \times 0.3 \text{ lbs/in.}^3 = 465 \text{ lbs} \quad (8.26)$$

(Assume entire weight is 500 lbs and is taken by the pipe support).

The total load on the soil is $2449 + 1029 + 500 = 3978$ lbs (8.27)

Then, the 16″ × 16″ or 1.78 ft² footing has a safety factor of

$$4000 \times 1.78/3978 = 1.79 \quad (8.28)$$

The length of the pipe is approximately

$$90'' - (1/2 \text{ the length of the duct depth of } 24'') = 78 \text{ in.} = 6.5 \text{ ft} \quad (8.29)$$

From the AISC handbook for columns, pages 3–34, $K = 1.2$, $L = 6.5$, then $KL = 7.8$ and for a 4″ standard pipe, the allowable load = 54 000 lbs.

The total axial load on the pipe is

$$F_V + \text{Duct weight} = 2449 + 500 = 2949 \text{ lbs} \quad (8.30)$$

The safety factor is

$$54\,000/2949 = 18.3 \quad (8.31)$$

9

System Explosion Isolation

The system explosion isolation must take into account the risk to personnel and equipment from the effects of the fireball temperature and the pressure. The hazard distances are calculated by the following equation:

$$D = \text{K\_flame}^{*}(V/N)^{1/3} \text{ (refer NFPA 68 – 2007)} \tag{9.1}$$

where, Distance D = Extension of fireball from the vent opening, m,

K_flame* = Flame length factor = 10 for metal dusts and 8 for chemical and agricultural dusts.

V = Volume of the dust collector, m^3

$N = 1$ = Number of explosion vents on unit

For illustration, from Figure 2.1, the total volume of the filter section is $98.75 \times 38.25 \times 57.59 = 217188 \text{ in.}^3$ (9.2)

$V = 217\,188 \text{ in.}^3 \times 1.639 \times 10^{-5} \text{ m/in.}^2 = 3.569 \text{ m}^3$

Then,

$$D = 8(3.569/1)^{1/3} = 8(1.528) = 12.22 \text{ m} \times 3.281 \text{ ft/m} = 40.1 \text{ ft diameter} \tag{9.3}$$

The radius of the projected fireball measured from the centerline of the vent shall be calculated as

$$1/2D = 40.1/2 = 20.05 \text{ ft} \tag{9.4}$$

The edge of the fireball must be at least 7 ft above the ground or roof to clear any personnel passing underneath. The edge of the fireball must also be clear of any access ladders, catwalks, doors, and vulnerable equipment. Avoid any air intake installations in the area (Figure 9.1).

Explosion Vented Equipment System Protection Guide, First Edition. Robert C. Comer.

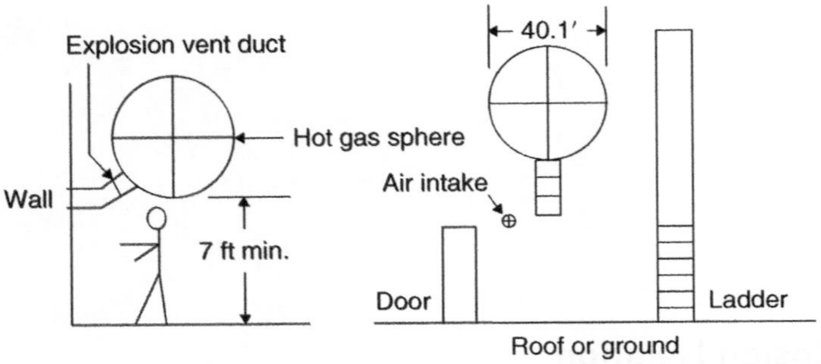

Figure 9.1 Explosion vent surrounding area exposure.

Pressure Profile

The maximum estimated overpressure from the vented explosion is

$$P_{\max\_a} = 0.2(P_{\text{Red}})(A_{\text{v}})^{0.1}(V)^{0.18} \text{ (refer NFPA 68} - 2007) \tag{9.5}$$

$0.3\ \text{m}^3 \leq V \leq 10\,000\ \text{m}^3$

$P_{\text{Red}} \leq 1$ bar

$P_{\text{stat}} \leq 0.1$ bar

$P_{\max} \leq 9$ bar

$K_{\text{st}} \leq 200$ bar-m/s

$P_{\max\_a}$ = External pressure (bar)

P_{Red} = Reduced pressure (bar) = 5.0 psi/14.5 = 0.345 bar

$$A_{\text{v}} = \text{Vent area } (\text{m}^2) = 24 \times 24 = 576\ \text{in.}^2/144\ \text{in.}^2/\text{ft}^2 = 4\ \text{ft}^2 \times 0.0928\ \text{m}^2/\text{ft}^2 = 0.372\ \text{m}^2 \tag{9.6}$$

V = Enclosure volume = 3.569 m^3

$$P_{\max\_a} = 0.2(0.345)(0.372)^{0.1}(3.569)^{0.18} = 0.079\ \text{bar} \times 14.5\ \text{psi/bar} = 1.14\ \text{psi} \tag{9.7}$$

Solving this equation for radius, r, when $P_{\max\_r} = 1.0$ psi (0.06896 bar)

$$r = P_{\max\_a}, (0.2)(D/0.06896), \text{where } D = 12.22\ \text{m and } P_{\max\_a} = 0.079\ \text{bar} \tag{9.8}$$

$$r = 2.80\ \text{m} \times 3.281\ \text{ft/m} = 9.2\ \text{ft} \tag{9.9}$$

The diameter of the fireball where the pressure is

$$1.0\ \text{psi is } 9.2 \times 2 = 18.4\ \text{ft} \tag{9.10}$$

10

Screw Conveyors, Rotary Airlock Valves, and Isolation Valves

Screw conveyors located on the discharge of a dust collector have the potential of allowing a dust explosion flame to pass through it downstream to other equipment. To preclude this from happening, it is recommended that part of the screw is removed, or a rotary airlock valve is employed on the inlet to the screw conveyor. Remove one-and-one half of a flight of the screw midway along the length of the screw conveyor. Install a baffle plate that blocks the upper half of the screw. Attach baffle to the cover. This ensures that a plug of the powder or dust remains as a choke. If a rotary airlock is located on the inlet to the screw conveyor, it is not necessary to form a choke (Figures 10.1 and 10.2).

The screw conveyor cover must resist the explosion flowing pressure, P_{Red}, unless a rotary airlock valve is placed at the inlet of the screw conveyor. Either bolts or clamps must be used to restrain the cover (Figures 10.3 and 10.4).

Assume that the cover is 0.1345″ thick × 10″ wide × 48″ long

$$a/b = 48/10 = 4.8 \text{ from Table 2.1}, \beta_1 = 0.500$$

$$f_{max} = \beta_1 P_{Red} b^2/t_p{}^2 = 0.500(5.0)\left(10^2\right)/0.1345^2 = 13\,819 \text{ psi}$$

Factor of safety = 21 440/13 819 = 1.55. The cover does not need reinforcing.

The separating load on the cover is $a \times b \times P_{Red} = 48 \times 10 \times 5.0 = 2400$ lbs.

Assume that there are eight bolts or clamps holding the cover on. The load per bolt is 2400/8 = 300 lbs/bolt. The bolts or clamps are not heavily loaded.

Rotary Airlock Valves

A minimum of eight vanes is required to ensure continuous sealing contact of the vanes with the housing to preclude a flame front from passing through the valve.

Rubber or plastic rotor tips will not withstand the heat of a deflagration. Metal tips are to be used.

Explosion Vented Equipment System Protection Guide, First Edition. Robert C. Comer.

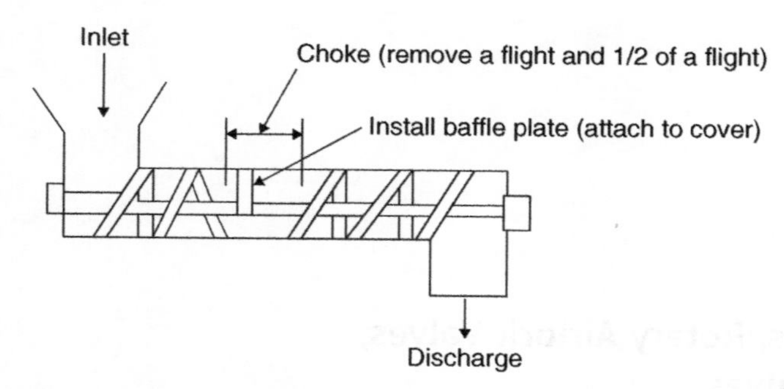

Figure 10.1 Screw conveyor choke.

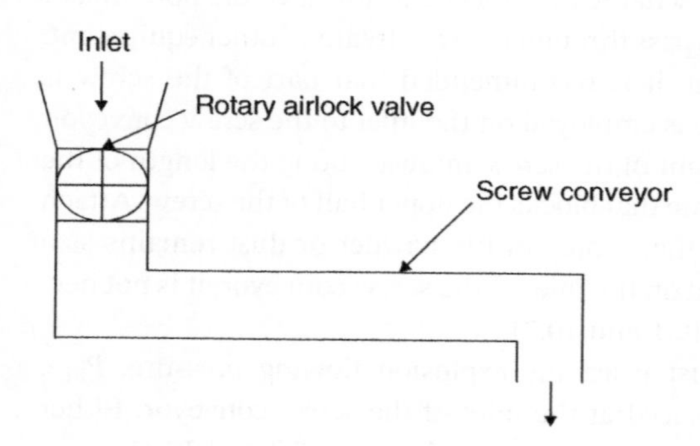

Figure 10.2 Screw conveyor with rotary airlock valve.

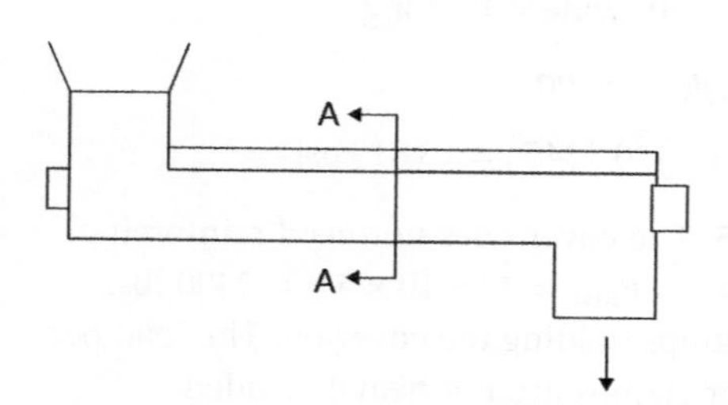

Figure 10.3 Screw conveyor cover.

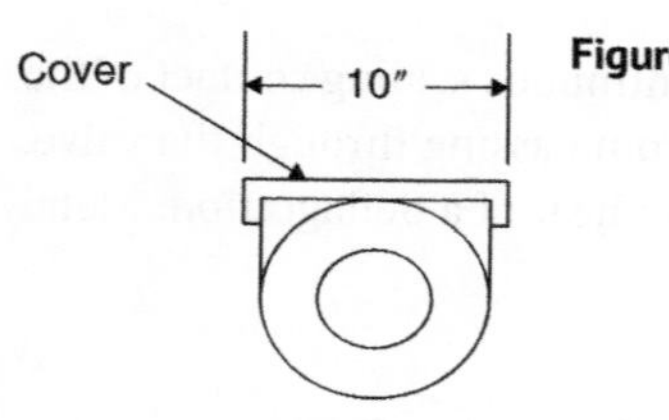

Figure 10.4 Section A–A.

Rotor vane to housing clearances must stay below 0.0079″ to maintain a seal of the hot gasses from passing through the valve. This is the valve manufacturer's recommendation.

Isolation Valves

Install an explosion isolation flap valve on the inlet duct to a dust collector. This self-actuating device will mitigate any explosion from propagating to upstream equipment. This device must be located at least 5 ft from the inlet and preferably as close to 10 ft as possible. This provides the time delay of the valve closing to preclude hot gas or particles from passing through the valve.

General

Wire the explosion vent's burst sensor to shut down the system when it is activated by an explosion.

Install a magnet above the inlet to any mill in the system to remove tramp metal that could cause a spark in the system.

Install a vibration switch on any mill in the system. Calibrate the switch to shut down the mill when the normal amplitude of vibration increases by 25% for more than one minute.

Install bearing temperature sensors on rotating equipment. Alarm the bearing temperature sensor to shut down the equipment or system when the normal steady-state reading increases by 30 °C.

11

Grounding of Systems

Grounding of all system elements is necessary to preclude inadvertent ignition of a dust cloud. The dust collector filter bags must be grounded as shown in Figure 11.1 if the bags are not inherently grounded. Electrically bond and ground all filter holders in the dust collector. Specify bags with 12″–18″ long braided copper wire sewn into the tops of bags. When installing the bags, loop the copper wire between the tube sheet, cage, and clamp. If the metal cages are not electrically bonded to the tube sheet, they act like capacitors. The venturi (nozzle) must be made of metal, not plastic. It provides the conductive path to the tube sheet.

There are anti-static bags that dissipate the static electricity to the nozzle; however, their disadvantage could be their filtering characteristics. Discuss with bag suppliers.

In many dust collectors, there is a metal grid near the bottom used for the operator to stand on while changing bags, etc. This grid must be electrically grounded or it could act like a capacitor.

Replace all carbon steel bolts, nuts, and washers with 304 stainless steel equivalents. This prevents nonconductive rust from forming at contact points.

Remove paint from all bolt and nut contact surfaces to provide for electrical bonding across bolted flanges.

Where bonded grounding system is all metal, resistance in continuous ground paths should be less than 10 Ω.

Operators must be grounded where any flammable atmosphere or explosive dust cloud with a minimum ignition energy less than 30 mj exists.

Use proper rated electrical equipment in locations where explosive atmospheres are or may be present. Conduct a hazardous area classification according to article 500 of the National Fire Protection Association (NFPA 70).

Flanged connection bonding is shown in Figure 11.2.

Morris coupling connection bonding is shown in Figure 11.3.

Flexible connection grounding between grounded components is shown in Figure 11.4.

Explosion Vented Equipment System Protection Guide, First Edition. Robert C. Comer.

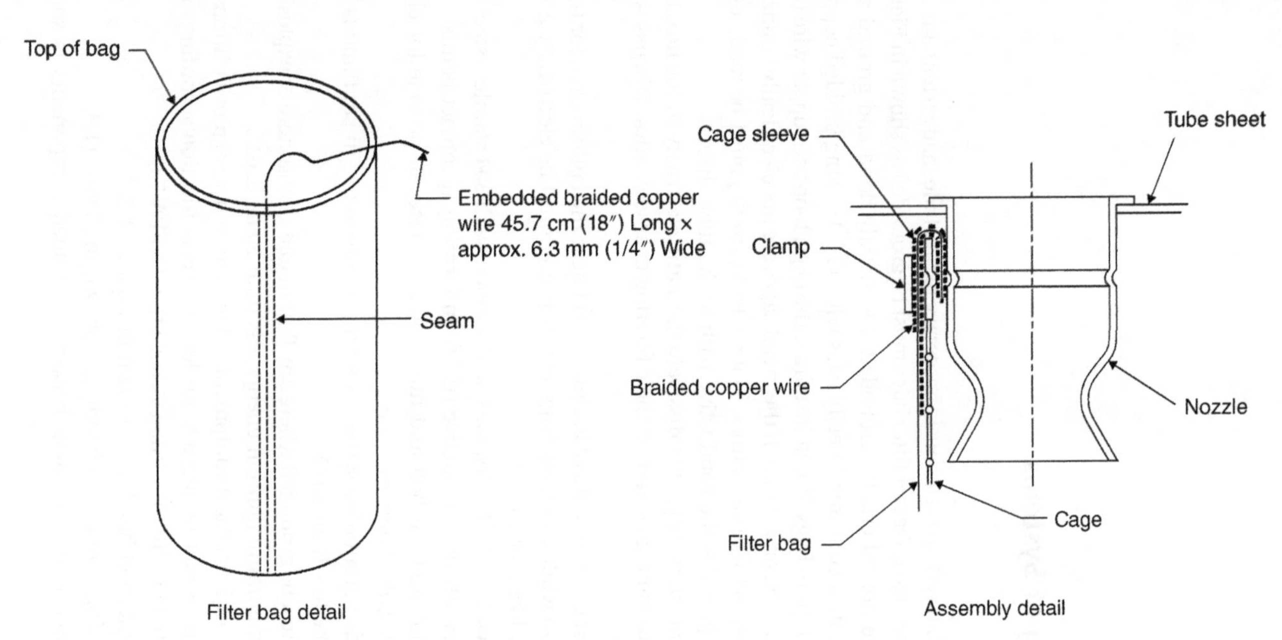

Figure 11.1 Grounded filter bag and cage assembly design detail.

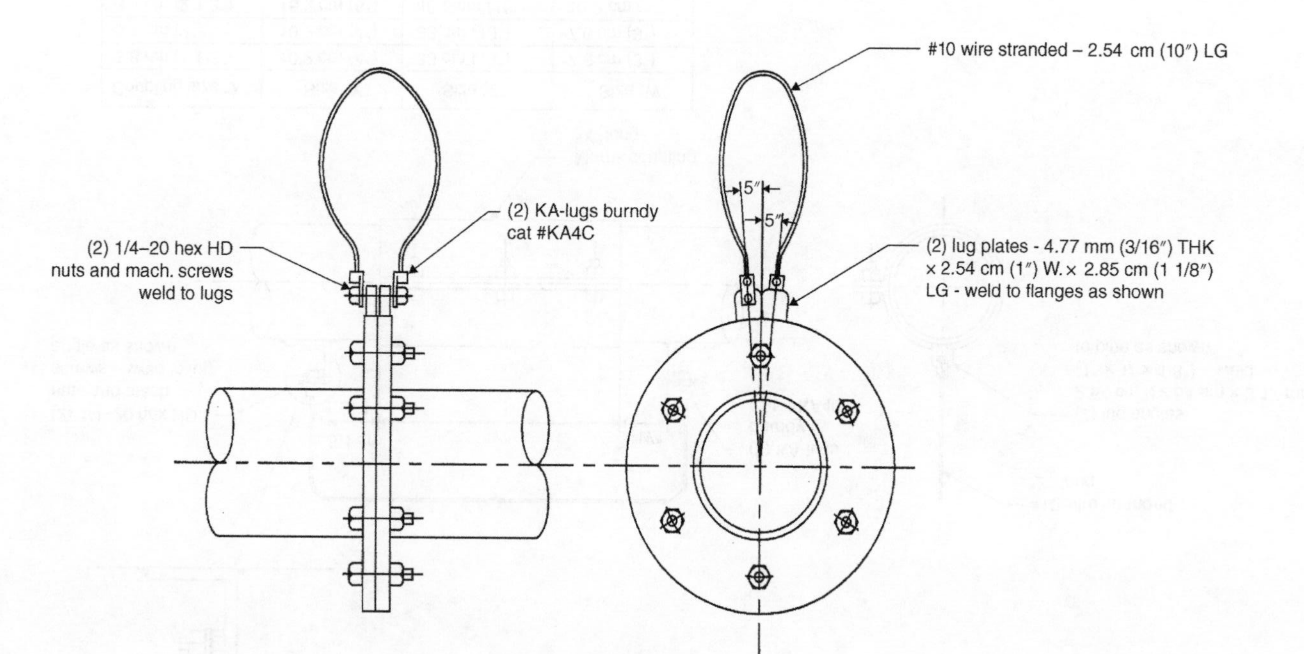

Figure 11.2 Flanged connection bonding and grounding details.

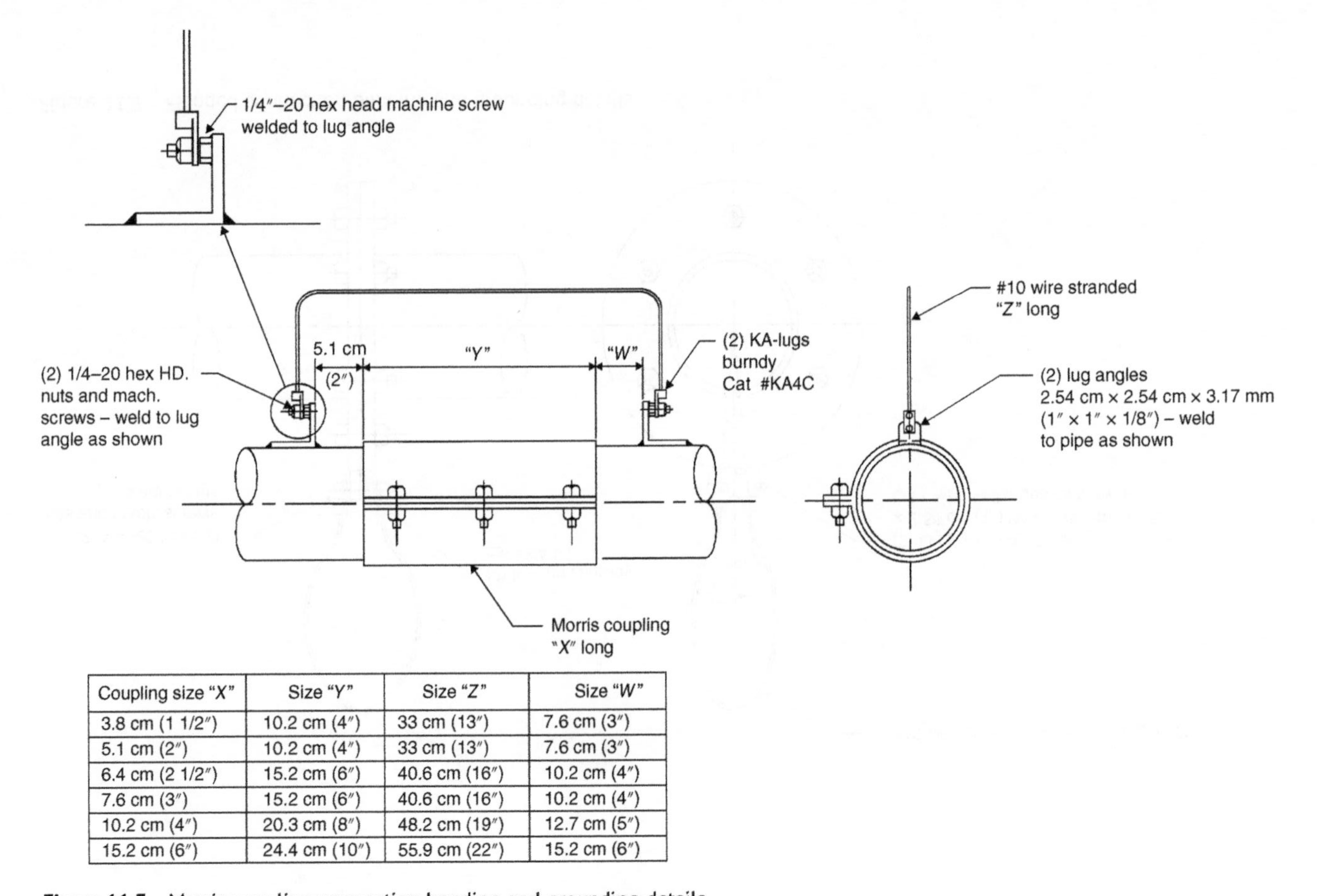

Coupling size "X"	Size "Y"	Size "Z"	Size "W"
3.8 cm (1 1/2″)	10.2 cm (4″)	33 cm (13″)	7.6 cm (3″)
5.1 cm (2″)	10.2 cm (4″)	33 cm (13″)	7.6 cm (3″)
6.4 cm (2 1/2″)	15.2 cm (6″)	40.6 cm (16″)	10.2 cm (4″)
7.6 cm (3″)	15.2 cm (6″)	40.6 cm (16″)	10.2 cm (4″)
10.2 cm (4″)	20.3 cm (8″)	48.2 cm (19″)	12.7 cm (5″)
15.2 cm (6″)	24.4 cm (10″)	55.9 cm (22″)	15.2 cm (6″)

Figure 11.3 Morris coupling connection bonding and grounding details.

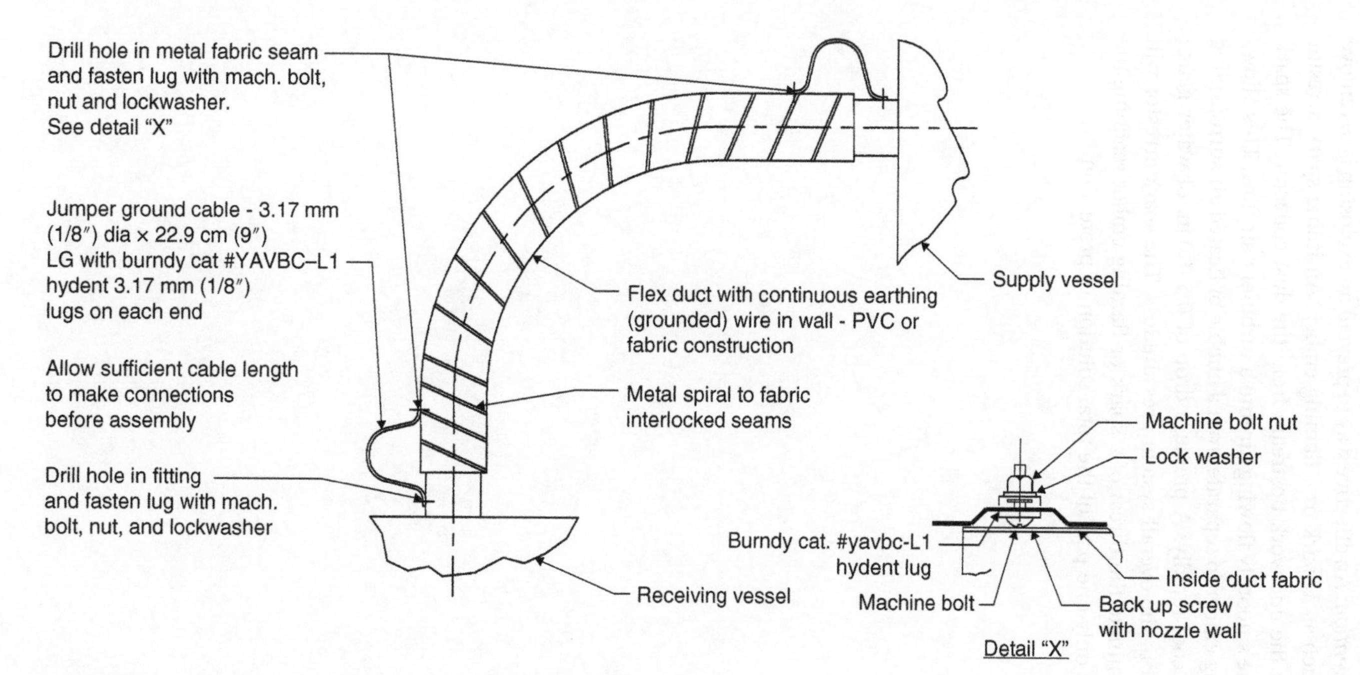

Figure 11.4 Flexible connection between grounded components detail.

Explosion ignition prevention: An effective way to prevent an explosion is to remove ignition sources, such as a spark or a flaming ember. An inline spark arrestor can be mounted in the ductwork upstream from the dust collector. The spark arrestor disrupts the smoothly flowing air into a turbulent air flow. This allows the spark or flaming ember to separate from a bubble of heated air surrounding it and allows it to cool rapidly. A pressure drop of 0.5–1.0 in. of water gauge should be included in the overall system pressure loss. The spark arrestor will not completely eliminate the threat of a spark or flaming ember reaching the dust collector. It is only one part of the total solution of protection.

12

Housekeeping and General Information

General housekeeping is an important part of keeping a facility safe from dust explosions. Prevent hazardous dusts from accumulating on surfaces. The standard allowable duct layer accumulation is 1/32 in. on at least 5% of the floor area. Floor area should be calculated to include dust on beams, ducts, equipment, and any vertical surfaces that dust sticks to. Maintaining a clean safe facility requires frequent cleaning of all surfaces in the area of dust handling. Compressed air blowing of dust is forbidden as an explosive dust cloud could be formed. Vacuuming is the preferred method of cleaning; however, industrial vacuum cleaners must be approved for use in Class 11, Division 2 areas. The vacuum should be a grounded stainless steel container with anti-static hose. Brushing or broom cleaning is not to be used. The dust could become airborne and spread to surfaces that are difficult to find and clean.

General Information

Perform welding, cutting, and any other hot work operations under a hot work permit procedure according to National Fire Protection Association (NFPA) 51B. Surfaces must be clean of dust prior to any hot work.

Mount an explosion vent area warning sign outside as shown in Figure 12.1 adjacent to the explosion vent duct in a highly visible area. It is important to notify any personnel in the area about the potential flame and heat hazard.

Explosion Vented Equipment System Protection Guide, First Edition. Robert C. Comer.

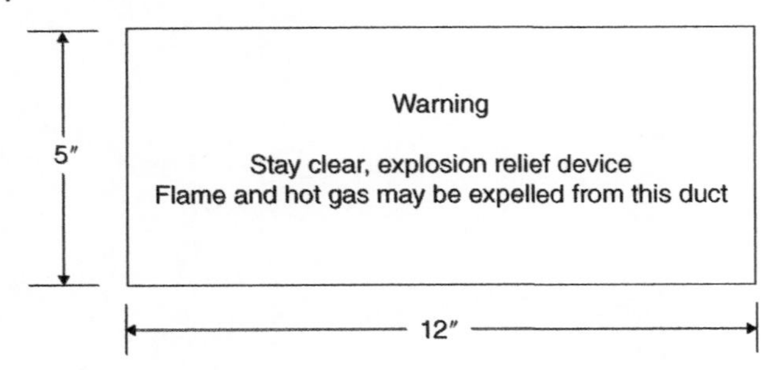

Figure 12.1 Explosion area warning sign.

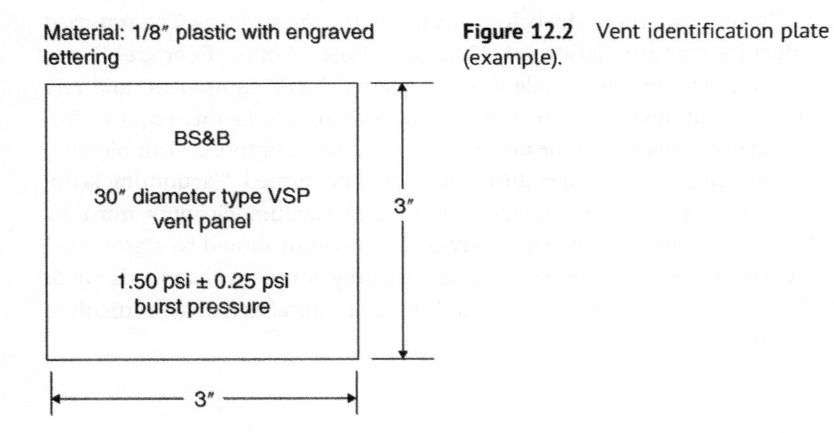

Figure 12.2 Vent identification plate (example).

Install a vent identification plate on the vent duct near the explosion vent to identify the burst element manufacturer, size, and burst pressure as shown in Figure 12.2. Installing the wrong pressure relief burst element for that system could be catastrophic.

Appendix A Part 1: Worksheet

Structural Tubing – Panel Worksheet (Example)

Composite Section of Tube with Panel

Properties of the tubular shapes are from AISC "Manual of Steel Construction."

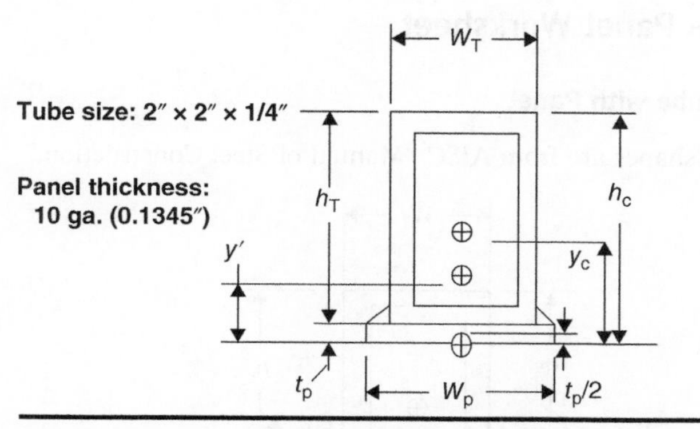

$W_p = W_T + 0.50 = 2.0 + 0.5 = 2.5''$ $A_p = W_p t_p = 2.5(0.1345) = 0.336$ sq in

$h_c = h_T + t_p = 2.1345''$ $y_c = h_T/2 + t_p = 1.1345$ $t_p/2 = 0.0670''$

Locate centroid axis with respect to base line: y'

Member	A_T	A_p	Y_c	$t_p/2$	Ay	$At_p/2$
Tube	1.59		1.1345		1.803	
Panel		0.336		0.067		0.022
Summary	1.926		—		1.825	

$Y' = Ay/A = 1.825/1.926 = 0.948$	$y_T = y_c - y' = 1.1345 - 0.948 = 0.186$	$y_p = y' - t_p/2 = 0.948 - 0.067 = 0.881$

Y = Distance of Center of gravity (C.G.) of section to C.G. of member.

I_o = Moment of inertia of a member about its own centroidal axis, for panel = $W_p t_p^3/12$ =

Member	A_T	A_p	Y_T	y_p	Y_T^2	y_p^2	Ay_T^2	Ay_p^2	I_o
Tube	1.59		0.186		0.035		0.055		0.766
Panel		0.336		0.948		0.899	0.302		0.0005
Summary	—		—		—		0.357		0.7665

$I_c = I_{combined}$ = Moment of inertia about centroidal axis of combined section at base, in.^{4}

$I_c = I_{combined}$, in.$^4 = Ay^2 + I_o = 0.357 + 0.766 = 1.124$ in.^{4}

S_c = Section modulus combined, in.$^3 = I_c/y' = 1.124/0.948 = 1.184$ in.^{3}

Structural Tubing – Panel Worksheet

Composite Section of Tube with Panel

Properties of the tubular shapes are from AISC "Manual of Steel Construction."

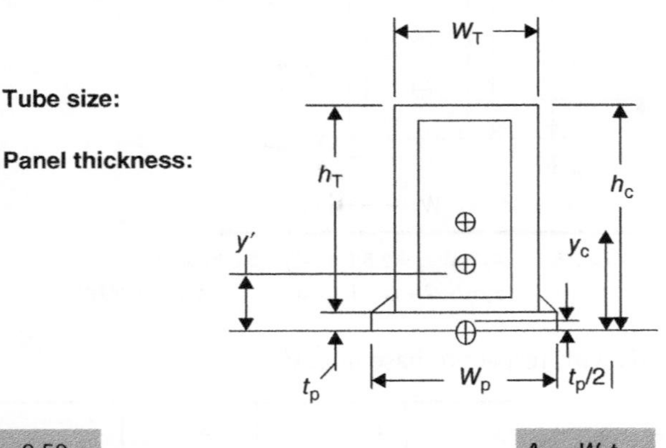

$W_p = W_T + 0.50 =$

$A_p = W_p t_p =$

$h_c = h_T + t_p =$

$y_c = h_T/2 + t_p =$

$t_p/2 =$

Locate centroid axis with respect to base line: y'

Member	A_A	A_p	Y_C	$t_p/2$	Ay	$At_p/2$
Tube						
Panel						
Summary			———			

$Y' = Ay/A=$	$y_T = y_c - y' =$	$y_p = y' - t_p/2=$

Y = Distance of C.G. of section to C.G. of member.

I_o = Moment of inertia of a member about its own centroidal axis, for panel = $W_p t_p^3/12 =$

Member	A_T	A_p	Y_T	y_p	Y_T^2	y_p^2	Ay_T^2	Ay_p^2	I_o
Tube									
Panel	—		—		—				
Summary	---------------		---------------		---------------				

$I_c = I_{combined}$ = Moment of inertia about centroidal axis of combined section at base, in.^{4}

$I_c = I_{combined}$, in.$^4 = Ay^2 + I_o =$

S_c = section modulus combined, in.$^3 = I_c/y' =$

Structural Angle – Panel Worksheet (Example)

Composite Section of Angle with Panel

Properties of the angle shapes are from AISC "Manual of Steel Construction."

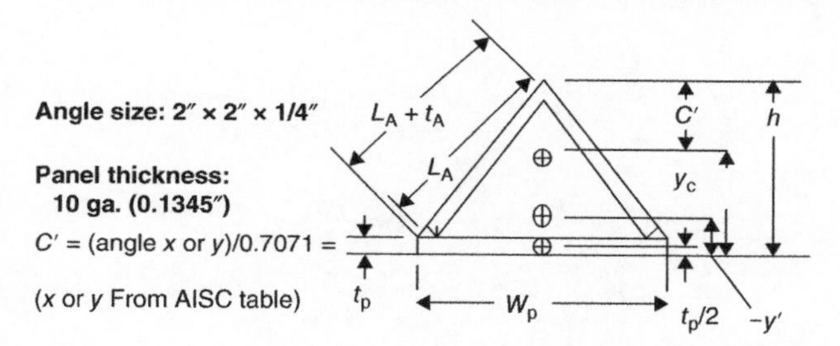

$W_p = (L_A + t_A)1.414 = 2.25(1.414) = 3.182$ $\quad A_p = W_p t_p = 3.182(0.1345) = 0.428$

$h = 0.7071(L_A + t_A) + t_p = 0.707(2.25) + 0.1345 = 1.725$

$y_c = h - C' = 1.725 - 0.837 = 0.888$ $\quad t_p/2 = 0.1345/2 = 0.067$

Locate centroidal axis with respect to base line: y'

Member	A_A	A_p	Y_C	$t_p/2$	Ay_c	$At_p/2$
Angle	0.938		0.888		0.833	
Panel		0.428		0.067		0.029
Summary	1.366		—		0.862	

$Y' = Ay/A = 0.862/1.366 = 0.631$	$y_A = y_c - Y' = 0.888 - 0.631 = 0.257$	$y_p = Y' - t_p/2 = 0.631 - 0.067 = 0.564$

Y = Distance of C.G. of section to C.G. of member.

I_o = Moment of inertia of a member about its own centroidal axis, for panel $= W_p t_p^3/12 = \dfrac{3.182(0.1345)^3}{12} = 0.000\,64$

For angle, $I_o = I_z = I_x \times 1.298$ where I_x is obtained from AISC table.

Member	A_A	A_p	y_A	y_p	y_A^2	y_p^2	Ay_A^2	Ay_p^2	I_o
Tube	0.938		0.257		0.066		0.062		0.450
Panel		0.428		0.564		0.318		0.136	0.0006
Summary	—		—		—		0.198		0.450

$I_c = I_{combined}$ = Moment of inertia about centroidal axis of combined section at base, in^4

$I_c = I_{combined}$, $in.^4 = Ay^2 + I_o = 0.198 + 0.450 = 0.648/in^4$

S_c = Section modulus combined, $in.^3 = I_c/Y' = \dfrac{0.648}{0.631} = 1.027/in^3$

Structural Angle – Panel Worksheet

Composite Section of Angle with Panel

Properties of the angle shapes are from AISC "Manual of Steel Construction."

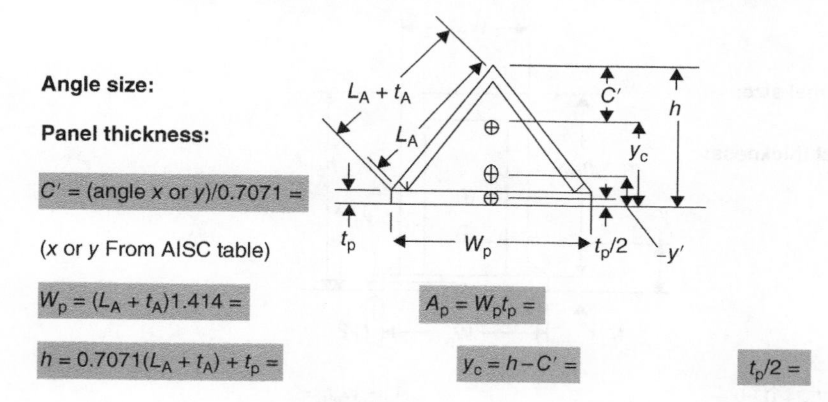

Angle size:

Panel thickness:

$C' = (\text{angle } x \text{ or } y)/0.7071 =$

(x or y From AISC table)

$W_p = (L_A + t_A)1.414 =$ $A_p = W_p t_p =$

$h = 0.7071(L_A + t_A) + t_p =$ $y_c = h - C' =$ $t_p/2 =$

Locate centroidal axis with respect to base line: y'

Member	A_A	A_p	Y_C	$t_p/2$	Ay_c	$At_p/2$
Angle						
Panel						
Summary			————			

$Y' = Ay/A=$	$y_A = y_c - Y' =$	$y_p = Y' - t_p/2=$

$Y =$ Distance of C.G. of section to C.G. of member.

$I_o =$ Moment of inertia of a member about its own centroidal axis, for panel $= W_p t_p^3/12 =$

For angle, $I_o = I_z = I_x \times 1.298 =$
where I_x is obtained from AISC table.

Member	A_A	A_p	y_A	y_p	y_A^2	y_p^2	Ay_A^2	Ay_p^2	I_o
Angle									
Panel									
Summary	---------------		---------------		---------------				

$I_c = I_{combined} =$ Moment of inertia about centroidal axis of combined section at base, in.^{4}

$I_c = I_{combined}$, in.$^4 = Ay^2 + I_o =$

$S_c =$ Section modulus combined, in.$^3 = I_c/Y' =$

Structural Channel – Panel Worksheet

Composite Section of Channel with Panel

Properties of the channels are from AISC "Manual of Steel Construction."

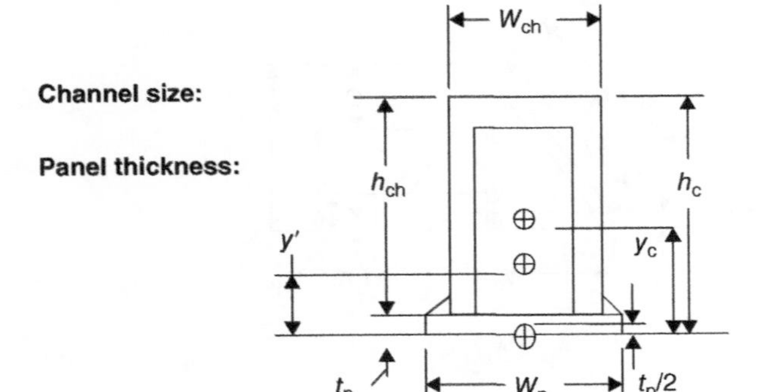

$W_p = W_{ch} + 0.50 =$ $A_p = W_p t_p =$

$h_c = h_{ch} + t_p =$ $y_c = h_{ch}/2 + t_p =$ $t_p/2 =$

Locate centroid axis with respect to base line: y'

Member	A_{ch} A_p	Y_c $t_p/2$	Ay $At_p/2$
Channel			
Panel			
Summary		————	

$Y' = Ay/A=$	$y_{ch} = y_c - Y' =$	$y_p = Y' - t_p/2=$

$Y =$ Distance of C.G. of section to C.G. of member.

$I_o =$ Moment of inertia of a member about its own centroidal axis, for panel $= W_p t_p^3/12 =$

Member	A_{ch} A_p	Y_{ch} y_p	Y_{ch}^2 y_p^2	AY_{ch}^2 Ay_p^2	I_o
Channel					
Panel					
Summary	---------------	---------------	---------------		

$I_c = I_{combined} =$ Moment of inertia about centroidal axis of combined section at base, in.^{4}

$I_c = I_{combined}$, in.^{4} $= Ay^2 + I_o =$

$S_c =$ section modulus combined, in.^{3} $= I_c/Y' =$

Reinforcing Rib – Panel Worksheet

Composite Section of Rib with Panel

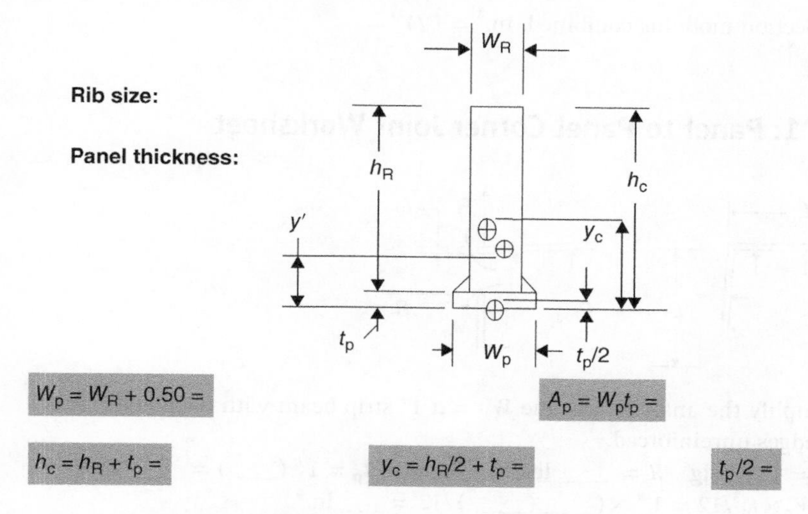

Locate centroid axis with respect to base line: y'

Member	A_R A_p	Y_c $t_p/2$	Ay $At_p/2$
Rib			
Panel			
Summary		———	

$Y' = Ay/A=$	$y_R = y_c - Y' =$	$y_p = Y' - t_p/2=$

$Y =$ Distance of C.G. of section to C.G. of member.

$I_o =$ Moment of inertia of a member about its own centroidal axis, for panel $= W_p t_p^3/12 =$

$I_o =$ Moment of inertia for rib $= W_R(h_R^3)/12 =$

Member	A_R A_p	Y_R y_p	Y_R^2 y_p^2	AY_R^2 Ay_p^2	I_o
Rib					
Panel					
Summary	---------------	---------------	---------------		

$I_c = I_{combined}$ = Moment of inertia about centroidal axis of combined section at base, in.^{4}

$I_c = I_{combined,in.^4} = Ay^2 + I_o =$

S_c = Section modulus combined, in.^{3} = $I_c/Y' =$

Detail 1: Panel to Panel Corner Joint Worksheet

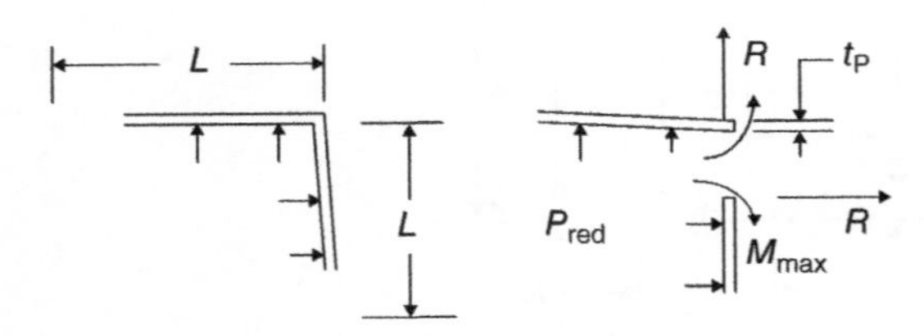

To simplify the analysis, assume $W_P = a$ 1″ strip beam with fixed ends. Panel corner edges unreinforced.

P_{RED} = ____psig R = _____lbs/in $A_p = W_p t_p$ = 1″ (____) = _____in.^{2}
$I_p = W_p \times t_p{}^3/12$ = 1″ × (____)(____^{3}) /12 = ____in.^{4}
$C = t_p/2$ = ____/2 = ____in.
$M_{max} = P_{Red} \times L^2/2$ = ____× (____^{2})/2 = _____in lbs
$f_{max} = R/A_p \pm M_{max}C/I_p$ = _____/____ ± ______ × _____/____ = + _____psi
This may be an extremely high stress in the unreinforced panel edges. If it is greater than 21 440 psi, reinforcing of the panel is required. Reinforce the corners with the main member angles or tubes. Refer to Section A.8.

Detail 2: Panel to Panel Corner Joint Reinforced Worksheet

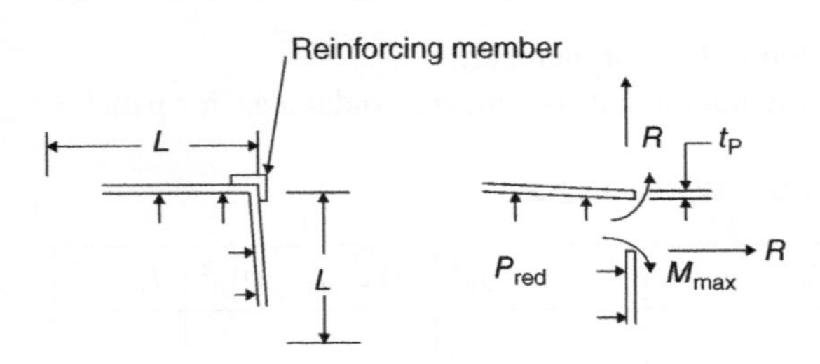

To simplify the analysis, assume $W_P = a\ 1''$ strip beam with fixed ends. Panel corner edges reinforced with a _____ × _____ × _____ angle

t_r = reinforcing member leg thickness, in.

P_{Red} = ____psig R = _____lbs/in

$A_p = W_p(t_p + t_r) = 1''$ (____+____) = _____in.^{2}

$I_p = W_p \times t_p^3/12 = 1''$ ×(____)(____^{3})/ 12 = ____in.^{4}

$C = (t_p + t_r)/2$ = (____+____)/2 = ____in.

$M_{max} = P_{Red} \times L^2/2$ = _____×(_____^{2})/2 = _____in lbs.

$f_{max} = R/A_p \pm M_{max}C/I_p$ = _____/____±______×_____/_____ = +_____psi.

Factor of safety = 28 800/_____=

Reinforce the corners to provide a safe structure.

Detail 3: Reinforcing Member Miter Joint Weld Worksheet

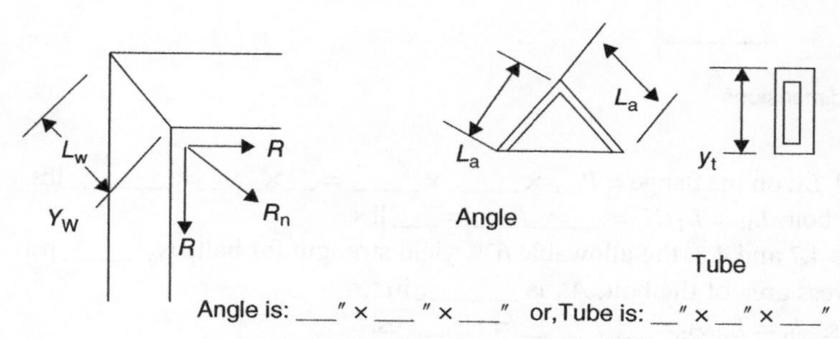

R = _____lbs, $R_n = R\sqrt{2}$ = ___(1.414) = _____lbs

For the angle:

L_a = Angle leg length plus the weld size = _____" + _____" = _____"

$2L_W = 2L_a$ for the angle = 2(___$\sqrt{2}$) = (_____1.414) = ___in.

From Table 4.1 for _____" × 1.5" pair of welds, F_a = allowable load = ______ lbs.

Factor of safety for the angle $= F_a(2L_W/1.5)/R_n$ = 28 800(______/1.5)/____ = ____

For the tube:

$Y_W = y_t\sqrt{2}$ = ___ × 1.414 = ___in.

$2Y_W$ = 2 weld lengths for the tube = 2×____ = ____in.

From Table 4.1 for _______" × 1.5" pair of welds, F_a = allowable load = ______ lbs.

Factor of safety for the tube $= F_a(2Y_W/1.5)/R_n$ = 28 800(____/1.5)/____ = ____

Detail 4: Square/Rectangular Bolted Flange Worksheet

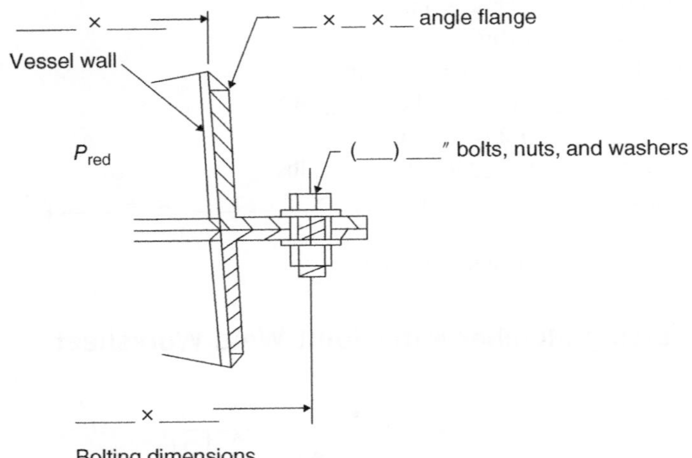

The total load, L_T, on the flange = P_{Red} × _____ × _____ = __ × ____ = _______ lbs.
The load per bolt, $L_B = L_T/N_B$= _____/___ = ___ lbs.
From Figures 4.7 and 4.8, the allowable 67% yield strength for bolts is _______ psi and the stress area of the bolt, A_s, is _______ in.^{2}.
The bolt stress, $f_b = L_B/A_S$= ___/_____ = _____ psi
Factor of safety, FS = allowable stress/bolt stress = ________/_______ + _____
Perimeter of flange = 2(_____) + 2(_____) = ___ in.

Square/rectangular bolted flange stress:

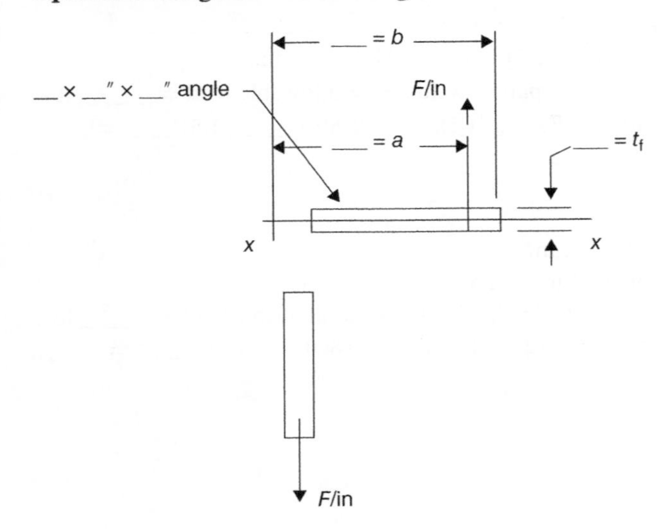

F/inch = L_T/perimeter of flange= ______/___ = ____lbs/in.
$I_{x-x} = bt_f^3/12$ = ___(___^{3})/12 = ____in.^{4}
$M_t = F$/inch $\times a$ = _____ × _____ = ______in-lbs in.
$C = t_f/2$ = ____/2 = ____in.
Stress = M_tC/I_{x-x} = ______ × ____/____ = ____psi
Factor of safety = 24 120/stress = 24 120/______

Detail 5: Round Duct Bolted Flange Worksheet

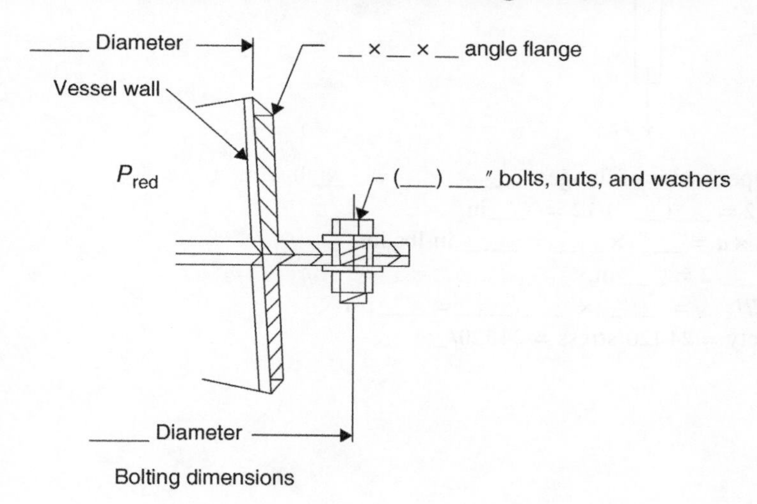

The total load, L_T, on the flange = $P_{Red}(\pi \times \text{diameter}^2)$ = ____ × (______^{2}) = ______lbs
The load per bolt, $L_B = L_T/N_B$ = ______/_____ = ______lbs/bolt
From Figures 4.7 and 4.8, the allowable 67% yield strength for bolts is _______ psi and the stress area of the bolt, A_s, is ________in.^{2}
The bolt stress, $f_b = L_B/A_S$= ___/______ = ______psi
Factor of safety, F.S. = allowable stress/bolt stress = _________/_______ + ______
Perimeter of flange = π(diameter) = ______in.

Round duct bolted flange stress:

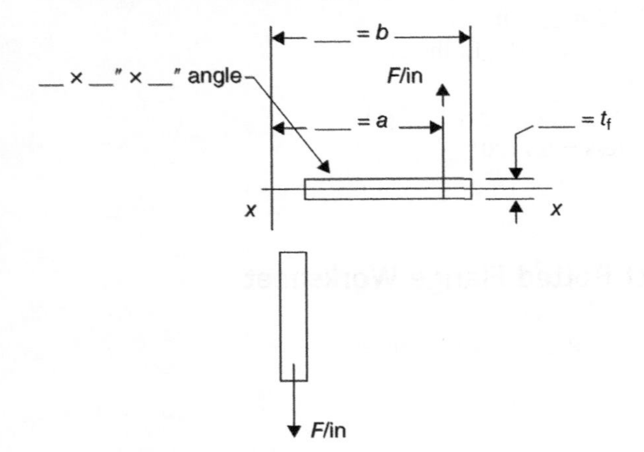

F/inch = L_T/perimeter of flange= ______/___ = ____ lbs/in.
$I_{x-x} = bt_f^3/12$ = ___(___^{3})/12 = _____in.^{4}
$M_t = F$/inch $\times a$ = ____ × ____ = ______in-lbs in.
$C = t_f/2$ = ____/2 = ____in.
Stress = $M_t C / I_{x-x}$ = _____ × ____/____ = ____psi
Factor of safety = 24 120/stress = 24 120/_____.

Part 2

Explosion Relief Element and Explosion Flowing Pressure Analyses

Part 2: Introduction

Explosion relief vent and explosion flowing reduced pressure (P_{Red}) analysis.

This section of the book is presented to allow a competent engineer that by education or experience understands the concepts presented in this book, to determine the explosion relief element to safely vent and lower the explosion flowing reduced pressure, P_{Red}, to a reasonable level so that reinforcing, if required, can be analyzed and designed in a practical, economical manner as defined in Part 1.

Required for this section is the process dust explosion characteristics, K_{st}, P_{max}, and the dust handling system geometry (refer to Part 2, Appendix B for a Dust Collector System Checklist). If the dust explosion characteristics are not known, a sample of the dust must be tested to obtain these characteristics. The table of dust characteristic values presented in Table 13.1 is for estimating purposes only. To ensure proper design a tested sample of dust must be used.

The characteristics of your dust may be quite different from the values listed due to particle size distribution, etc. and serious mistakes can be made by using the listed values without backing them up with tested values. NFPA 652, Appendix A, provides additional material lists, and the same caution must be used.

This Part 2 is also to be used if the product process or the product is changed from the product used to analyze the original venting of a system. If the process material is changed to a more highly explosive material, it may be necessary to reanalyze the equipment to ensure that the explosion vent and reinforcing is adequate. The difference in dust characteristics may be significant and the required reinforcing may be more extensive or a larger explosion vent is required.

Part 2 provided analysis for existing dust handling systems with explosion relief burst elements already sized and attached. It also provided analysis of new dust handling systems to ensure that the explosion relief burst element is properly sized during the fabrication process.

Explosion Vented Equipment System Protection Guide, First Edition. Robert C. Comer.

13

Know Your Process Dust Characteristics

Table 13.1 lists the generic characteristics (K_{st} and P_{max}) for various dusts. These values may not be the same as your process dust and are presented as guidance only. Your process dust must be tested in accordance with ASTM Standard E1226, "Standard Test Method for Pressure and Rate of Pressure Rise for Combustible Dusts."

The explosive behavior of dusts is described in terms of the maximum explosion pressure and rate of pressure rise.

K_{st} is the normalized rate of pressure rise of explosion pressure and is the average of the largest observed rates from each series of tests. P_{max} is the average of the largest explosion pressures observed from each of three test series.

The potential severity of a dust explosion is quantified by measured tested parameters and they are a function of the following:

Particle size distribution and chemical composition of the dust.
Concentration of the dust in the dust/air mixture.
Homogeneity and turbulence of the dust/air mixture.
Type, energy, and location of the ignition source.
Geometry of the vessel in which the explosion is taking place.
Initial pressure and temperature of the dust/air mixture.

Table 13.1 is compiled from various sources, primarily from National Fire Protection Association (NFPA)-68, "Standard on Explosion Protection by Deflagration Venting," and the dust characteristics are listed for guidance only. The values listed may not be the same as your process dust characteristics and **your process dust must be tested in accordance with ASTM Standard E1226, "Standard Test Method for Pressure and Rate of pressure Rise for combustible Dusts."** The potential severity of a dust explosion may be much higher than the listed values.

Explosion Vented Equipment System Protection Guide, First Edition. Robert C. Comer.

Table 13.1 Explosion characteristics of various dusts.

Material	K_{st} (bar-m/s)	P_{max} (bar)
Agriculture products		
Barley	50	8.0
Cellulose	229	9.7
Cork	202	9.6
Corn	75	9.4
Dried blood plasma	119	6.8
Egg white	38	8.3
Milk, powdered	28	5.8
Milk, nonfat, dry	125	8.8
Pasta	50	8.0
Peas	107	7.2
Potatoes	50	8.0
Soy flour	110	9.2
Starch, corn	202	10.3
Starch, rice	101	9.2
Starch, wheat	115	9.9
Sugar	138	8.5
Sugar, milk	82	8.3
Sugar, beet	59	8.2
Tapioca	62	9.4
Wheat flour	127	7.3
Whey	140	9.8
Wood flour	205	10.5
Carbonaceous dusts		
Charcoal, activated	14	7.7
Charcoal, wood	10	9.0
Coal, bituminous	129	9.2
Coke, petroleum	47	7.6
Lampblack	121	8.4
Lignite	151	10.0
Peat, 22% H_2O	67	84.0
Soot, pine	26	7.9

Table 13.1 (Continued)

Material	K_{st} (bar-m/s)	P_{max} (bar)
Metal dusts		
Aluminum	415	12.4
Bronze	31	4.1
Iron carbonyl	111	6.1
Magnesium	508	17.5
Niobium	238	6.3
Phenolic resin	269	7.9
Silicon	126	10.2
Tantalum	149	6.0
Zinc	176	7.3
Chemical dusts		
Adipic acid	97	8.0
Anthraquinone	364	10.6
Ascorbic acid	111	9.0
Calcium acetate	21	6.5
Calcium stearate	132	9.1
Carboxymethyl cellulose	136	9.2
Dextrin	106	8.8
Lactose	81	7.7
Lead stearate	152	9.2
Methyl cellulose	134	9.5
Paraformaldehyde	178	9.9
Sodium ascorbate	119	8.4
Sodium stearate	123	8.8
Sulfur	151	6.8
Plastic dusts		
There is an extended list of plastic dusts in NFPA-68, Table E.1(e).		

Source: Data from NFPA 68, "Standard on Explosion Protection by Deflagration Venting"; ASTM Standard E1226, "Standard Test Method for Pressure and Rate of pressure Rise for combustible Dusts"

14

Venting Analysis of Dust Handling Systems

This chapter is in accordance with NFPA 68 (2013).

Square/Rectangular Dust Collector

This chapter presents the analysis of dust handling systems that are in service or are to be ordered and have an explosion vent installed. Existing systems may or may not have documentation of any prior analysis and/or there may have been a change in the process dust being handled. The example used for the square/rectangular dust collector is the example used and illustrated in Part 1, Figure 2.1. Figures 14.1–14.3 schematically illustrates the vessel and hopper parameters and the effective volumes to be used in the venting analyses. NFPA 68 (2013) "Standard on Explosion Protection by Deflagration Venting" is used for all analyses in this part.

As per NFPA 68 (2013) Para. 6.4.3.3.1, "Internal volume of dust collector bags, filters, or cartridges shall be permitted to be eliminated when determining the effective volume of an elongated enclosure, when the vent is positioned as required by 8.7.1(1) or 8.7.1(2)." This requires the explosion vent to be located below the filter bag areas. In almost every case, the filter bags, etc., are extended below the explosion vent location even for a vent located near the hopper, as the manufacturer uses as much filter area as possible. Therefore, for these analyses, there is no deduction for the filter bag volume in the calculations. This approach also provides a greater safety factor on venting. If, in the event that your system places the explosion vent below the filter bags, it is a simple filter bag volume deduction from the volume calculations. Table 14.1 presents conversion factors.

Explosion Vented Equipment System Protection Guide, First Edition. Robert C. Comer.

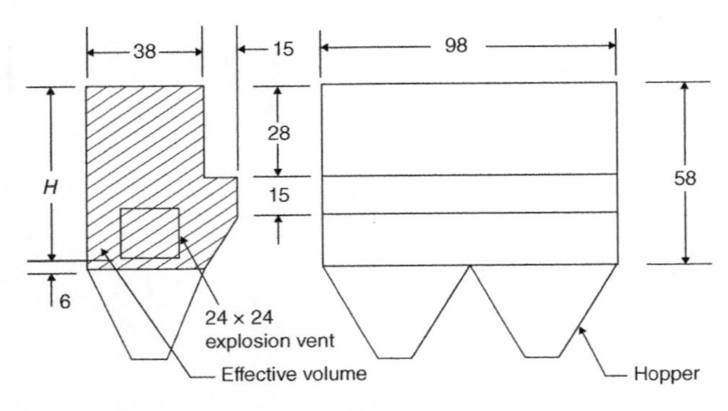

Figure 14.1 Square/rectangular dust collector with explosion vent close to hopper.

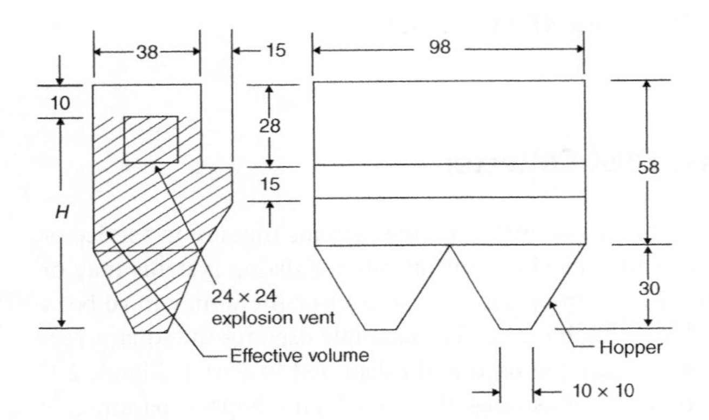

Figure 14.2 Square/rectangular dust collector with explosion vent high on the vessel.

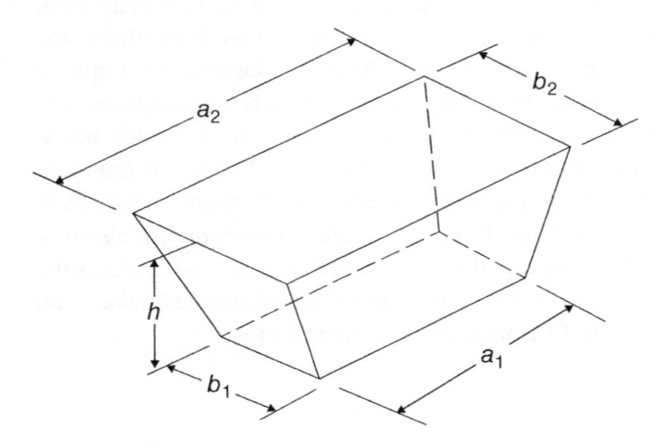

Figure 14.3 Square/rectangular hopper volume.

Table 14.1 Conversion factors.

To convert	Into	Multiply by
Bars	Pounds/square inches	14.5
Cubic feet	Cubic inches	1728
Cubic feet	Cubic meters	0.028 32
Cubic inches	Cubic feet	5.787×10^{4}
Cubic inches	Cubic meters	1.639×10^{-5}
Cubic meters	Cubic feet	35.31
Cubic meters	Cubic inches	61 023
Feet	Meters	0.3048
Inches	Meters	2.540×10^{-2}
Inches of mercury	Pounds/square inches	0.4912
Inches of water	Pounds/square inches	0.036 13
Meters	Feet	3.281
Meters	Inches	39.37
Pounds/cubic feet	Pounds/cubic inches	5.787×10^{-4}
Pounds/cubic inches	Pounds/cubic feet	1728
Pounds/square feet	Inches of mercury	0.014 14
Pounds/square feet	Pounds/square inches	6.944×10^{-3}
Pounds/square inches	Inches of mercury	2.036
Pounds/square inches	Pounds/square feet	144
Square feet	Square inches	144
Square feet	Square meters	0.092 90
Square inches	Square feet	6.944×10^{-3}
Square meters	Square feet	10.76
Square meters	Square inches	1550

Example with Figure 14.1: Existing Vessel with Explosion Vent Close to Hopper

For this example, it is assumed that there is a change from the old process dust to a new process corn starch dust, and the existing vessel has a 24″ × 24″ explosion vent installed. A new P_{Red} for the vessel must be calculated and compared to the prior reinforcing for a P_{Red} of 5.0 psi that was performed in Part 1. If the new P_{Red} exceeds 5.0 psi, then the reinforcing performed in Part 1 must be reanalyzed or a larger vent must be installed to bring the explosion flowing pressure down to a safe level.

Symbols: Refer to NFPA 68 (2013)

A_{vo} = Vent area (m^2)
P_{stat} = Nominal static burst pressure of the vent (bar)
P_{max} = Maximum pressure of a deflagration (bar)
P_{Red} = Reduced pressure after deflagration venting (bar)
K_{st} = Deflagration index (bar-m/s)
V_{eff} = Enclosure volume (m^3). Volume from the top of the vessel to the bottom of the vent.

$$A_{vo} = 1 \times 10^{-4}\left(1 + 1.54P_{stat}^{4/3}\right)K_{st}V_{eff}^{3/4}\sqrt{(P_{max}/P_{Red}) - 1} \text{ per NFPA (2013) Para. 8.2.2} \quad (14.1)$$

where

$$A_{vo} = 24 \text{ in.} \times 24 \text{ in.} = 576 \text{ in.}^2 = \frac{4 \text{ ft}^2}{10.76 \text{ ft}^2/\text{m}^2} = 0.372 \text{ m}^2 \quad (14.2)$$

$$P_{stat} = \frac{1.50 \text{ psi}}{14.5 \text{ psi/bar}} = 0.103 \text{ bar} \quad (14.3)$$

P_{max} = 10.3 bar (from Table 13.1) (your corn starch dust must be tested to establish the actual P_{max})

P_{Red} = A value to be calculated (psi)

K_{st} = 202 (bar-m/s) (from Table 13.1) (your corn starch dust must be tested to establish the actual K_{st})

$$V_{eff} = (38 \times 58 - 6) + (15 \times 15) + (58 - 28 - 15) \times 15/2 = 2313 \text{ in.}^2 \times 98 \text{ in. long}$$
$$= 226\,674 \text{ in.}^3 \times 1.639 \times 10^{-5} \text{ m}^3/\text{in.}^3 = 3.72 \text{ m}^3 \quad (14.4)$$

With the explosion vent close to the hopper, the hopper volume is not included.

$$L/D = H/D_{he} \quad (14.5)$$

$$L = H = 58 - 6 = 52 \text{ in.} \times 2.540 \times 10^{-2} \text{ m/in.} = 1.32 \text{ m} \quad (14.6)$$

$$D_{he} = 4 \times A_{eff}/p \quad (14.7)$$

$$A_{eff} = V_{eff}/H = 3.72/1.32 = 2.82 \text{ m}^2 \quad (14.8)$$

$$p = 2 \times (38 + 98) = 272 \text{ in.} \times 2.540 \times 10^{-2} = 6.91 \text{ m} = \text{perimeter of vessel} \quad (14.9)$$

$$D_{he} = 4 \times 2.82/6.91 = 1.63\ \text{m} \tag{14.10}$$

$$L/D = 1.32/1.63 = 0.81 \tag{14.11}$$

$L/D \leq 2.0$ therefore, no enclosure correction is required.

In accordance with NFPA 68 (2013), Para. 8.2.2.3, "When L/D is less than or equal to 2, A_{vi} shall be set equal to A_{vo}."

NFPA 68 (2013), Para. 8.2.3 "For L/D values greater than 2 and less than or equal to 6, the required vent area, A_{vi}, shall be corrected and calculated as follows:

$$A_{vi} = A_{vo}\left[1 + 0.6(L/D - 2)^{0.75} \times \exp\left(-0.95 \times {P_{Red}}^2\right)\right] \tag{14.12}$$

where

$$\exp(A) = e^A, \quad \text{e is the base of the natural logarithm} \tag{14.13}$$

Substituting values and solving for P_{Red} with new product corn starch dust and existing explosion vent.

$$0.372 = 1 \times 10^{-4}\left(1 + 1.54 \times 0.103^{4/3}\right)202 \times 3.72^{3/4}\sqrt{(10.3/P_{Red}) - 1}$$

$$0.372 = 1 \times 10^{-4}(1.074)541\sqrt{(10.3/P_{Red}) - 1}$$

$$0.372 = 0.0581\sqrt{(10.3/P_{Red}) - 1}$$

$$6.40 = \sqrt{(10.3/P_{Red}) - 1}$$

$$1 + 6.40^2 = 10.3/P_{Red} \tag{14.14}$$

$$P_{Red} = 10.3/41.96 = 0.245\ \text{bar} \times 14.5\ \text{psi/bar} = 3.55\ \text{psi} \tag{14.15}$$

This pressure is lower than the 5.0 psi; therefore, the example vessel is safe for corn starch use as reinforced in Part 1.

The process corn starch dust must be tested to determine the actual dust explosion characteristics.

To check the calculations substituting values and solving for vent area V_{vo} with $P_{Red} = 3.55\ \text{psi} = 0.245\ \text{bar}$

$$\begin{aligned} V_{vo} &= 1 \times 10^{-4}\left(1 + 1.54 \times 0.103^{4/3}\right)202 \times 3.72^{3/4}\sqrt{(10.3/0.245) - 1} \\ &= 1 \times 10^{-4}(1.074)541\sqrt{10.3/0.245 - 1} \\ &= 0.0581\sqrt{(10.3/0.245) - 1} \\ &= 0.0581 \times 6.40 = 0.372\ \text{m}^2 \times 1550\ \text{in.}^2/\text{m}^2 = 577\ \text{in.}^2 \end{aligned} \tag{14.16}$$

The explosion vent size required to reduce the explosion flowing pressure P_{Red} to 3.55 psi is approximately $\sqrt{577} = 24'' \times 24''$ square. The vent is $24'' \times 24''$, so the calculations check out. The vessel is safe for corn starch use.

Refer to Chapter 15 **for the effect on the explosion vent size by duct back pressure and other considerations.**

Example with Figure 14.2: Vessel with Explosion Vent High on Vessel

For this example, it is assumed that the new process is corn starch and the existing vessel has a $24'' \times 24''$ explosion vent installed. A new P_{Red} for the vessel must be calculated and compared to the prior reinforcing for a P_{Red} of 5.0 psi that was performed in Part 1. If the new P_{Red} exceeds 5.0 psi, then the reinforcing performed in Part 1 must be reanalyzed or a larger vent must be installed to bring the explosion flowing pressure down to a safe level.

The hopper volume must be included in this example.

$$A_{vo} = 24\text{ in.} \times 24\text{ in.} = 576\text{ in.}^2 = \frac{4\text{ ft}^2}{10.76\text{ ft}^2/\text{m}^2} = 0.372\text{ m}^2 \tag{14.17}$$

$$P_{stat} = \frac{1.50\text{ psi}}{14.5\text{ psi/bar}} = 0.103\text{ bar} \tag{14.18}$$

$P_{max} = 10.3$ bar (from Table 13.1) (your corn starch dust must be tested to establish the actual P_{max})

P_{Red} = A value to be calculated (psi)

$K_{st} = 202$ (bar-m/s) (from Table 13.1) (your corn starch dust must be tested to establish the actual K_{st})

$$\begin{aligned} V_{eff} &= (38 \times 58 - 10) + (15 \times 15) + (58 - 28 - 15) \times 15/2 = 2161\text{ in.}^2 \times 98\text{ in. long} \\ &= 211\,778\text{ in.}^3 \times 1.639 \times 10^{-5}\text{ m}^3/\text{in.}^3 = 3.47\text{ m}^3 \end{aligned} \tag{14.19}$$

Hopper volume: Refer to Figure 14.4 and NFPA 68 (2013), Figure A.6.4.3(e)

$$\begin{aligned} V_h = {} & [a_1 \times h \times (b_2 - b_1)/2] + [b_1 \times h \times (a_2 - a_1)/2] \\ & + [h \times (a_2 - a_1) \times (b_2 - b_1)/3] + a_1 \times b_1 \times h \end{aligned} \tag{14.20}$$

where

$a_1 = 10$ in.

$a_2 = 49$ in.

$b_1 = 10$ in.

$b_2 = 38$ in.

$h = 30$ in.

$$V_h = [10 \times 30 \times (38-10)/2] + [10 \times 30 \times (49-10)/2] + [30 \times (49-10) \times (38-10)/3] + 10 \times 10 \times 30$$
$$= 4200 + 5850 + 10\,920 + 3000 = 23\,970 \text{ in.}^3 \times 1.639 \times 10^{-5} = 0.393 \text{ m}^3$$

$$\text{For two hoppers} = 0.393 \times 2 = 0.786 \text{ m}^3 \quad (14.21)$$

$$V_{\text{eff}} = \text{The total effective volume} = 3.47 + 0.786 = 4.26 \text{ m}^3 \quad (14.22)$$

$$L/D = H/D_{\text{he}} \quad (14.23)$$

$$L = H = 58 + 30 - 10 = 78 \text{ in.} \times 2.540 \times 10^{-2} \text{ m/in.} = 1.98 \text{ m} \quad (14.24)$$

$$D_{\text{he}} = 4 \times A_{\text{eff}}/p \quad (14.25)$$

$$A_{\text{eff}} = V_{\text{eff}}/H = 4.26/1.98 = 2.15 \text{ m}^2 \quad (14.26)$$

$$p = 2 \times (38 + 98) = 272 \text{ in.} \times 2.540 \times 10^{-2} = 6.91 \text{ m} = \text{perimeter of vessel} \quad (14.27)$$

$$D_{\text{he}} = 4 \times 2.15/6.91 = 1.24 \text{ m} \quad (14.28)$$

$$L/D = 1.98/1.24 = 1.60 \quad (14.29)$$

$L/D < 2.0$; therefore, no L/D correction is necessary.

Substituting values and solving for P_{Red} with product corn starch dust and existing explosion vent.

$$0.372 = 1 \times 10^{-4}\left(1 + 1.54 \times 0.103^{4/3}\right)202 \times 4.26^{3/4}\sqrt{(10.3/P_{\text{Red}}) - 1}$$
$$0.372 = 1 \times 10^{-4}(1.074)599\sqrt{(10.3/P_{\text{Red}}) - 1}$$
$$0.372 = 0.064\sqrt{(10.3/P_{\text{Red}}) - 1}$$
$$5.81 = \sqrt{(10.3/P_{\text{Red}}) - 1}$$

$$1 + 5.81^2 = 10.3/P_{\text{Red}} \quad (14.30)$$

$$P_{\text{Red}} = 10.3/34.76 = 0.296 \text{ bar} \times 14.5 \text{ psi/bar} = 4.29 \text{ psi} \quad (14.31)$$

The result of this analysis is approximately the same as the prior analysis. The vessel as reinforced in Part 1 for $P_{\text{Red}} = 5.0$ psi is safe for corn starch use.

The process corn starch dust must be tested to determine the actual explosion characteristics.

Refer to Chapter 15 for the effect on the explosion vent size by duct back pressure and other considerations.

For vessels with multiple explosion vents, refer to NFPA 68 (2013), Para. A.6.4.3.

Cylindrical Dust Collector

Figures 14.4–14.6 schematically illustrates the vessel and hopper parameters and the effective volumes to be used in the venting analysis. NFPA 68 (2013) "Standard on Explosion Protection by Deflagration Venting" is used for all analyses in this part.

Example with Figure 14.4: Cylindrical Vessel with Explosion Vent on the Side

For this example, it is assumed that a new process dust is corn starch, and the existing vessel has a 24″ × 24″ explosion vent installed on the side of the vessel. A new P_{Red} for the vessel must be calculated and compared to the P_{Red} = 5.0 psi reinforcing on Figure 3.1.

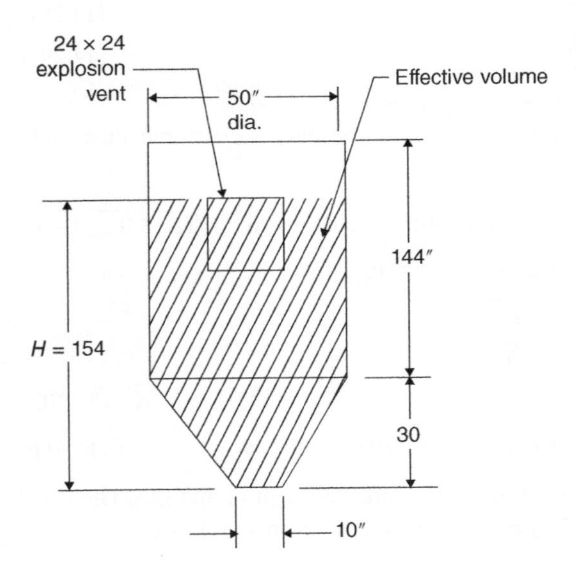

Figure 14.4 Cylindrical dust collector with an explosion vent on the side of the vessel.

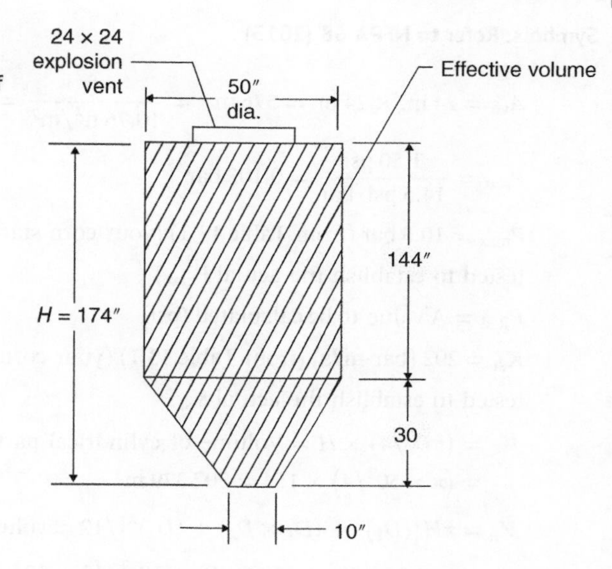

Figure 14.5 Cylindrical dust collector with an explosion vent on the top of the vessel.

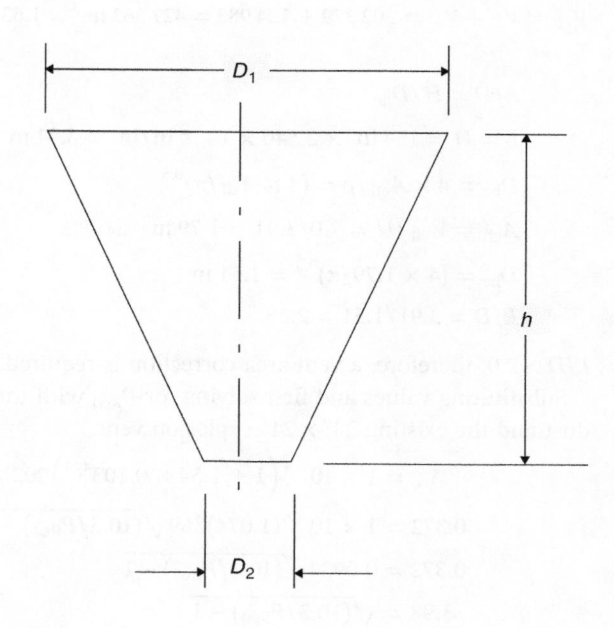

Figure 14.6 Cylindrical hopper volume.

Symbols: Refer to NFPA 68 (2013)

$$A_{vo} = 24\text{ in.} \times 24\text{ in.} = 576\text{ in.}^2 = \frac{4\text{ ft}^2}{10.76\text{ ft}^2/\text{m}^2} = 0.372\text{ m}^2 \tag{14.32}$$

$$P_{stat} = \frac{1.50\text{ psi}}{14.5\text{ psi/bar}} = 0.103\text{ bar} \tag{14.33}$$

$P_{max} = 10.3$ bar (from Table 13.1) (your corn starch dust must be tested to establish the actual P_{max})

P_{Red} = A value to be calculated (psi)

$K_{st} = 202$ (bar-m/s) (from Table 13.1) (your corn starch dust must be tested to establish the actual K_{st})

$$\begin{aligned} V_C &= (\pi D^2/4) \times H = \text{Volume of cylindrical part of vessel} \\ &= (\pi \times 50^2/4) \times 154 = 302\,379\text{ m}^3 \end{aligned} \tag{14.34}$$

$$\begin{aligned} V_h &= \pi H\left[(D_1)^2 + (D_1 \times D_2) + (D_2)^2\right]/12 = \text{volume of conical hopper} \\ &= \pi \times 154\left[50^2 + (50 \times 10) + 10^2\right]/12 = 124983\text{ in.}^3 \end{aligned} \tag{14.35}$$

$$V_{eff} = V_C + V_h = 302\,379 + 124\,983 = 427\,362\text{ in.}^3 \times 1.639 \times 10^{-5}\text{m}^3/\text{in.}^3 = 7.0\text{ m}^3 \tag{14.36}$$

$$L/D = H/D_{he} \tag{14.37}$$

$$L = H = 154\text{ in.} \times 2.540 \times 10^{-2}\text{ m/in.} = 3.91\text{ m} \tag{14.38}$$

$$D_{he} = 4 \times A_{eff}/p = (4 \times A_{eff}/\pi)^{0.5} \tag{14.39}$$

$$A_{eff} = V_{eff}/H = 7.0/3.91 = 1.79\text{ m}^2 \tag{14.40}$$

$$D_{he} = (4 \times 1.79/\pi)^{0.5} = 1.51\text{ m} \tag{14.41}$$

$$L/D = 3.91/1.51 = 2.58 \tag{14.42}$$

$L/D > 2.0$; therefore, a vent area correction is required.

Substituting values and first solving for P_{Red} with the new process corn starch dust and the existing 24″ × 24″ explosion vent.

$$A_{vo} = 1 \times 10^{-4}\left(1 + 1.54 \times 0.103^{4/3}\right)202 \times 7.0^{3/4}\sqrt{(10.3/P_{Red}) - 1}$$

$$0.372 = 1 \times 10^{-4}(1.074)869\sqrt{(10.3/P_{Red}) - 1}$$

$$0.372 = 0.0934\sqrt{(10.3/P_{Red}) - 1}$$

$$3.98 = \sqrt{(10.3/P_{Red}) - 1}$$

$$1 + 3.98^2 = 10.3/P_{Red} \tag{14.43}$$

$$P_{Red} = 10.3/16.84 = 0.611\text{ bar} \times 14.5\text{ psi/bar} = 8.87\text{ psi} \tag{14.44}$$

8.87 psi exceeds the $P_{Red} = 5.0$ psi that the example vessels in Part 1 were reinforced to. It is more economical to replace the 24″ × 24″ explosion vent with a larger vent.

Calculate the required explosion vent size required to bring P_{Red} down to 5.0 psi (0.345 bar).

$$A_{vo} = 1 \times 10^{-4}\left(1 + 1.54 \times 0.103^{4/3}\right)202 \times 7.0^{3/4}\sqrt{(10.3/P_{Red}) - 1}$$
$$= 1 \times 10^{-4}(1.074)869\sqrt{(10.3/0.345) - 1}$$
$$= 0.0933\sqrt{(10.3/0.345) - 1} \quad (14.45)$$

$$A_{vo} = 0.0933 \times 5.372 = 0.501\ \text{m}^2 \times 1550\ \text{in.}^2/\text{m}^2 = 777\ \text{in.}^2 \quad (14.46)$$

The explosion vent size to bring down P_{Red} to 5.0 psi is approximately $\sqrt{777} = 28'' \times 28''$ uncorrected.

NFPA 68 (2013), Para. 8.2.3 For L/D values greater than 2 and less than or equal to 6, the required vent area, A_{v1}, shall be corrected and calculated as follows:

$$A_{vi} = A_{vo}\left[1 + 0.6(L/D - 2)^{0.75} \times \exp\left(-0.95 \times P_{Red}^2\right)\right] \quad (14.47)$$

where

$$\exp(A) = e^A, \quad \text{e is the base of the natural logarithm} \quad (14.48)$$

$$A_{vi} = 0.501\left[1 + 0.6(2.58)^{0.75} \times \exp\left(-0.95 \times 0.345^2\right)\right]$$
$$= 0.501[2.22 \times \exp(-0.113)]$$
$$= 0.501[2.22 \times 0.893] = 0.993\ \text{m}^2 \times 1550\ \text{in.}^2/\text{m}^2 = 1539\ \text{in.}^2 \quad (14.49)$$

The explosion vent size, corrected is approximately $\sqrt{1539} = 39'' \times 39''$.

It may be more practical to keep the 24″ width of the explosion vent and increase the height to 1539 in.²/24 in. = 64‴.

Another option is to make the explosion vent 36″ wide × 1539/36 = 43″ high.

Refer to Chapter 15 **for the effect on the explosion vent size by duct back pressure and other considerations.**

Example with Figure 14.5: Cylindrical Vessel with Explosion Vent on the Top

For this example, it is assumed that a new process dust is corn starch, and the existing vessel has a 24″ × 24″ explosion vent installed on the side of the vessel. A new P_{Red} for the vessel must be calculated and compared to the $P_{Red} = 5.0$ psi reinforcing on Figure 3.1.

Symbols: Refer to NFPA 68 (2013)

A_{vo} = Vent area (m^2) uncorrected
A_{vi} = Vent area (m^2) corrected for L/D greater than 2.0
P_{stat} = Nominal static burst pressure of the vent (bar)
P_{max} = Maximum pressure of a deflagration (bar)
P_{Red} = Reduced pressure after deflagration venting (bar)
K_{st} = Deflagration index (bar-m/s)
V_{eff} = Enclosure volume (m^3). Volume from the bottom of the hopper to the top of the vent.

$$A_{vo} = 24\text{ in.} \times 24\text{ in.} = 576\text{ in.}^2 = \frac{4\text{ ft}^2}{10.76\text{ ft}^2/\text{m}^2} = 0.372\text{ m}^2 \tag{14.50}$$

$$P_{stat} = \frac{1.50\text{ psi}}{14.5\text{ psi/bar}} = 0.103\text{ bar} \tag{14.51}$$

P_{max} = 10.3 bar (from Table 13.1) (your corn starch dust must be tested to establish the actual P_{max})

P_{Red} = A value to be calculated (psi)

K_{st} = 202 (bar-m/s) (from Table 13.1) (your corn starch dust must be tested to establish the actual K_{st})

$$\begin{aligned} V_C &= (\pi D^2/4) \times h = \text{Volume of cylindrical part of vessel} \\ &= (\pi \times 50^2/4) \times 144 = 282\,744\text{ in.}^3 \end{aligned} \tag{14.52}$$

$$\begin{aligned} V_h &= \pi h\left[(D_1)^2 + (D_1 \times D_2) + (D_2)^2\right]/12 = \text{Volume of conical hopper} \\ &= \pi \times 30\left[50^2 + (50 \times 10) + 10^2\right]/12 = 24347\text{ in.}^3 \end{aligned} \tag{14.53}$$

$$V_{eff} = V_C + V_h = 282\,744 + 24\,347 = 307\,091\text{ in.}^3 \times 1.639 \times 10^{-5}\text{m}^3/\text{in.}^3 = 5.03\text{ m}^3 \tag{14.54}$$

$$L/D = H/D_{he} \tag{14.55}$$

$$L = H = (144 + 30)\text{ in.} \times 2.540 \times 10^{-2}\text{ m/in.} = 4.42\text{ m} \tag{14.56}$$

$$D_{he} = 4 \times A_{eff}/p = (4 \times A_{eff}/\pi)^{0.5} \tag{14.57}$$

$$A_{eff} = V_{eff}/H = 5.03/3.91 = 1.29\text{ m}^2 \tag{14.58}$$

$$D_{he} = (4 \times 1.29/\pi)^{0.5} = 1.281\text{ m} \tag{14.59}$$

$$L/D = 4.42/1.281 = 3.45 \tag{14.60}$$

L/D is higher than 2.0; therefore, a vent area correction is required.

Substituting values and first solving for P_{Red} with the new process corn starch dust and the existing 24″ × 24″ explosion vent.

$$A_{vo} = 1 \times 10^{-4}\left(1 + 1.54 \times 0.103^{4/3}\right)202 \times 5.03^{3/4}\sqrt{(10.3/P_{Red}) - 1}$$

$$0.372 = 1 \times 10^{-4}(1.074)678\sqrt{(10.3/P_{Red}) - 1}$$

$$0.372 = 0.0728\sqrt{(10.3/P_{Red}) - 1}$$

$$5.11 = \sqrt{(10.3/P_{Red}) - 1}$$

$$1 + 5.11^2 = 10.3/P_{Red} \quad (14.61)$$

$$P_{Red} = 10.3/27.11 = 0.379\ \text{bar} \times 14.5\ \text{psi/bar} = 5.51\ \text{psi} \quad (14.62)$$

5.51 psi exceeds the $P_{Red} = 5.0$ psi that the example vessels in Part 1 were reinforced to. It is more economical to replace the 24″ × 24″ explosion vent with a larger vent.

Calculate the required explosion vent size required to bring P_{Red} down to 5.0 psi (0.345 bar).

$$A_{vo} = 1 \times 10^{-4}\left(1 + 1.54 \times 0.103^{4/3}\right)202 \times 5.03^{3/4}\sqrt{(10.3/P_{Red}) - 1}$$

$$= 1 \times 10^{-4}(1.074)678\sqrt{(10.3/0.345) - 1}$$

$$= 0.0728\sqrt{(10.3/0.345) - 1} \quad (14.63)$$

$$A_{vo} = 0.0728 \times 5.372 = 0.391\ \text{m}^2 \times 1550\ \text{in.}^2/\text{m}^2 = 606\ \text{in.}^2 \quad (14.64)$$

The explosion vent size to bring down P_{Red} to 5.0 psi is approximately $\sqrt{606} = 24.6'' \times 24.6''$ uncorrected.

NFPA 68 (2013), Para. 8.2.3 For L/D values greater than 2 and less than or equal to 6, the required vent area, A_{vi}, shall be corrected and calculated as follows:

$$A_{vi} = A_{vo}\left[1 + 0.6(L/D - 2)^{0.75} \times \exp\left(-0.95 \times P_{Red}^2\right)\right] \quad (14.65)$$

where

$$\exp(A) = e^A, \quad \text{e is the base of the natural logarithm} \quad (14.66)$$

$$A_{vl} = 0.391\left[1 + 0.6(3.45)^{0.75} \times \exp\left(-0.95 \times 0.345^2\right)\right]$$

$$= 0.391[2.52 \times \exp(-0.113)]$$

$$= 0.391[2.52 \times 0.893] = 0.880\ \text{m}^2 \times 1550\ \text{in.}^2/\text{m}^2 = 1363\ \text{in.}^2 \quad (14.67)$$

The explosion vent size, corrected is approximately $\sqrt{1363} = 37'' \times 37''$.

It may be more practical to keep the 24″ width of the existing explosion vent and increase the height to 1363 in.^{2}/24 in. = 57‴.

Refer to Chapter 15 **for the effect on the explosion vent size by duct back pressure and other considerations**.

15

Duct Back Pressure Considerations

The explosion vent sizing and the explosion vent duct analyses have not considered the duct back pressure effect and other considerations that can occur in the duct. Listed below are considerations that increase the explosion vent area. A significant rise in the explosion flowing reduced pressure, P_{Red}, may occur due to the friction in a long duct and the resistance to flow in a duct with bends.

This chapter is in accordance with NFPA 68 (2013). For additional detail, refer to NFPA 68 (2013).

The corrections to the explosion vent area are as follows:

A_{v1} for L/D for values greater than 2.0 (if no correction was required as calculated in Chapter 14) $A_{v1} = A_{vo}$

A_{v2} for when v_{axial} or v_{tan} is greater than 20 m/s (if both values are less than 20 m/s, A_{v2} shall be set equal to A_{v1}).

A_{v3} for when the vent panel mass, M, exceeds the threshold mass, M_T.

A_{v4} for when there is a partial volume deflagration (when partial volume is not applied, $A_{v4} = A_{v3}$).

From Figure 14.2: Vessel with explosion vent high on the vessel. The explosion vent size did not have to be corrected from 24″ × 24″ (576 in.^{2}) as L/D is less than 2.0. P_{Red} was calculated to be 4.29 psi (0.296 bar).

Using the previous explosion vent duct example from Figure 6.1, assume that the duct is 24″ × 24″, and the maximum duct length is 20 ft overall with a long sweep 90° elbow (bent in 22-1/2° increments) to keep resistance to flow to a minimum (Figure 15.1).

These considerations are:

A_{v1}: L/D is less than 2.0. No correction required.

A_{v2}: Per NFPA 68 (2013), "This correction applies when the average air axial velocity, v_{axial}, and the tangential velocity, v_{tan}, are both less than 20 m/s during all operating conditions." It was assumed that the velocities were less than 20 m/s without an engineering analysis to determine the velocities.

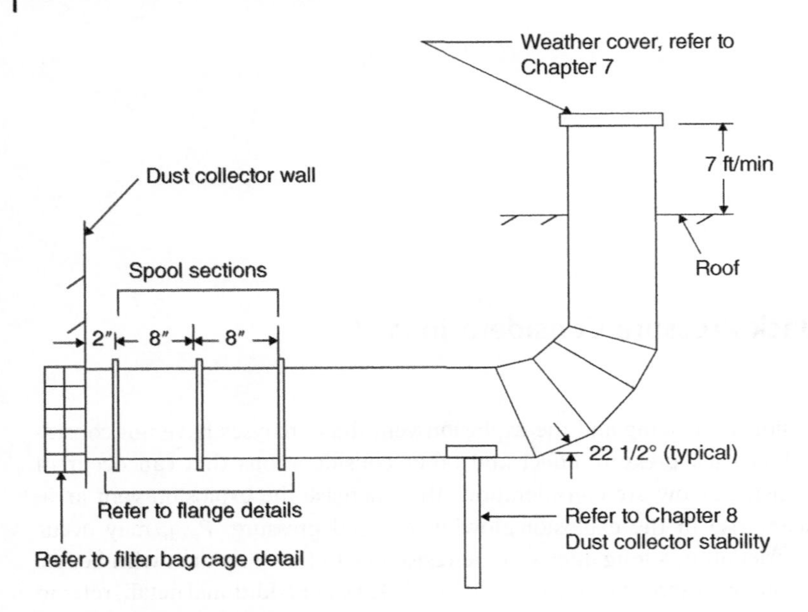

Figure 15.1 Explosion vent duct-roof.

If an engineering analysis is performed and the velocities are greater than 20 m/s, then a correction by A_{v2} is required as follows:

$$v_{\text{axial}} = Q_{\text{air}} \times L/V \text{ refer NFPA 68 (2013) (8.2.6.2)} \tag{15.1}$$

where

Q_{air} = Flow rate through the equipment (m^3/s)
L = Equipment overall length (m) in the direction of air and product flow
V = Equipment volume (m^3)

If a circumferential (tangential) air velocity is in the equipment, $v_{\tan}$ shall be given by $0.5v_{\tan\_\max}$ where $v_{\tan\_\max}$ is the maximum tangential air velocity in the equipment.

Per NFPA 68 (2013): Para. 8.2.6.4 "Values of Q_{air}, v_{axial}, $v_{\tan\_\max}$, and $v_{\tan}$ shall be either measured or calculated by engineers familiar with the equipment design and operation."

Per NFPA 68 (2013): Para. 8.2.6.6 "When the maximum value derived of v_{axial} and $v_{\tan}$ are less than 20 m/s, A_{v2} shall be set equal to A_{v1}.

When either v_{axial} or $v_{\tan}$ is larger than 20 m/s, A_{v2} shall be determined from the following equation:

$$A_{v2} = A_{v1}\left[1 + \frac{\max(v_{axial}, v_{tan}) - 20}{36} \times 0.7\right] \text{ refer NFPA 68 (2013) (8.2.5.7)} \tag{15.2}$$

where max(*A*, *B*) = maximum value of either *A* or *B*.

There is no calculation of velocities for this unit; therefore, assume both velocities are less than 20 m/s. Therefore, $A_{v2} = A_{v1}$

A_{v3}: Panel mass correction

Per NFPA 68 (2013) Para. 8.2.6, "When the mass of the vent panel is less than or equal to 40 kg/m² and K_{st} is less than or equal to 250 bar-m/s, NFPA 68 (2013) Eq. (8.2.6.2) shall be used to determine whether an incremental increase in vent area is needed."

$$M_T = \left[6.67\left(P_{Red}^{0.2}\right)\left(n^{0.3}\right)\left(V/K_{st}^{0.5}\right)\right]^{1.67} \quad 8.2.6.2 \tag{15.3}$$

where

M_T = Threshold mass (kg/m²)
P_{Red} = 4.29 psi (0.296 bar)
n = Number of panels (1)
V = Volume (m³) = 4.26
$K_{st} \leq 250$ bar-m/s (202)

$$M_T = \left[6.67\left(0.296^2\right)\left(1^{0.3}\right)\left(4.26/202^{0.5}\right)\right]^{1.67} = [0.175]^{1.67} = 0.055 \text{ kg/m}^2 \tag{15.4}$$

Per NFPA 68 (2013), Para. 8.2.7, If $M \geq M_T$, the vent area shall be increased by adding the calculated area, A_{v3} from Eq. (8.2.7):

$$A_{v3} = F_{sh}\left[1 + (0.0075) \times M^{0.6}\left(\frac{K_{st}^{0.5}}{n^{0.3} V P_{Red}^{0.2}}\right)\right] A_{v2} \text{ (8.2.7)} \tag{15.5}$$

where

A_{v2} = Vent area corrected for prior conditions (m²)
F_{sh} = 1.0 per NFPA 68 (2013)
M = Mass of vent panel (kg/m²) = 2.5 lbs/ft² × 4.882 = 12.2 kg/m²
(2.5 lbs/ft² is the approximate mass of the explosion relief panels from Fike or BSB)

Then

$$A_{v3} = 1 \times \left[1 + (0.0075) \times 12.2^{0.6}\left(\frac{202^{0.5}}{1^{0.3} \times 4.26 \times 0.296^{0.2}}\right)\right] A_{v2} \tag{15.6}$$
$$= 4.40 A_{v2} = 4.40 \times 0.372 = 1.636 \text{ m}^2 = 2537 \text{ in.}^2$$

Approximately a 50″ × 50″ explosion relief panel or a 36″ × 70″ panel.

A_{v4}: Partial volume correction

Per NFPA 68 (1013) Para. 8.3, When the volume fill fraction, X_v, can be determined for a worst-case explosion scenario, the minimum required vent area shall be permitted to be calculated from the following equation:

$$A_{v4} = A_{v3} \times X_v^{-1/3} \sqrt{(X_v - \Pi)/(1 - \Pi)} \quad (8.3.1) \qquad (15.7)$$

where

A_{v4} = Vent area for partial volume deflagration
A_{v3} = Vent area for full volume deflagration as determined from Eq.(8.2.7)
X_v = Fill fraction $\geq \Pi$
$\Pi = P_{red}/P_{max}$

If $X_v \leq \Pi$, deflagration venting shall not be required. When partial volume is not applied, $A_{v4} = A_{v3}$. No partial volume was applied in this example.

Duct Back Pressure Correction

The previous explosion vent sizing and the explosion vent duct analyses have not considered the back pressure effect that can occur in the duct. A significant rise in the explosion flowing reduced pressure (P_{Red}) may occur due to the friction in a long duct and the resistance to flow in a duct with bends.

NFPA 68 (2013), Para. 8.5 "Effects of vent ducts," and Para A.8.5 presents equations to determine the adjusted vent area for various vent ducts. Per NFPA 68 (2013), If a new P_{Red} is desired while keeping the explosion vent duct size at the original value of 24″ × 24″, The solution is iterative, as E_1 and E_2 are both functions of A_{vf}. In this example case, $P_{Red} = 4.29$ psi is desired and the resultant explosion vent duct size is adjusted for the correction factors while maintaining a P_{Red} less than 5.0 psi in order to maintain the reinforcing design in Part 1.

Using example explosion vent size of 36″ × 70″ after being corrected for the above considerations, calculate the required vent duct size to compensate for the duct friction and fitting resistance.

$$A_{vf} = A_{v3}\left(1 + 1.18 \times E_1^{0.8} \times E_2^{0.4}\right) \times \sqrt{K/K_o} \quad (8.5.1a) \qquad (15.8)$$

where

A_{vf} = Vent area required when a duct is attached to the vent opening (m^2)

A_{v3} = Vent area after adjustment for partial volume or in this case $A_{v3} = A_{v2}$ = 1.63 m^2

$$E_1 = \frac{A_{vf} \times L_{duct}}{V} \text{ (8.5.1b) constrain } A_{vf} \text{ to } \leq 1.0 \text{ per NFPA 68 (2013) Para. A.8.5} \tag{15.9}$$

$$E_2 = \frac{10^4 \times A_{vf}}{(1 + 1.54 \times P_{stat}{}^{4/3})K_{st} \times V^{3/4}} \text{ (8.5.1c) constrain } A_{vf} \text{ to } \leq 1.0 \text{ per NFPA 68 (2013) Para. A.8.5} \tag{15.10}$$

where

P_{stat} = Nominal static opening pressure of the vent cover 0.103 bar

V = Enclosure volume 4.26 m^3

L_{duct} = Vent duct overall length 20 ft × 0.3048 = 6.096 m

K_o = 1.5, the resistance coefficient value assumed that validated Eqs.(8.2.2) and (8.2.3)

$$K = \frac{\Delta P}{1/2\rho \times U^2} = K_{inlet} + \frac{f_D \times L}{D_h} + K_{elbows} + K_{outlet} + \cdots \text{ (8.5.1d)} \tag{15.11}$$

where

K_{inlet} = Resistance coefficient for fittings from Figure A.8.5a = 0.49

K_{outlet} = 0

W/D = Aspect ratio of duct = width/depth = 68/36 = 1.9 say 2.0

R/D = Bend radius/depth of duct = 54/36 = 1.5

K_{elbows} = Elbow loss coefficient from Figure A.8.5c for R/D = 1.5 and W/D = 2.0 = 0.13

K = Overall resistance coefficient of the vent duct application

U = Fluid velocity

D_h = Vent duct hydraulic diameter (m) = 4(2.82/6.91) = 1.63 m (refer Eq.(14.10))

f_D = D'Arcy friction factor for fully turbulent flow; (see NFPA 68 (2013) A.8.5a for typical formula)

e = Effective roughness = 0.25 mm for used ducts

$$f_D = \left\{\frac{1}{[1.14 - 2\log_{10}(e/D_h)]}\right\}^2 \quad (A.8.5a)$$

$$= \left\{\frac{1}{[1.14 - 2\log_{10}(0.26/1.63)]}\right\}^2 = 0.012$$

$$K = 0.49 + \frac{0.012 \times 6.096}{1.63} + 0.13 + 0 = 0.664$$

Then

$$E_1 = \frac{1.63 \times 6.096}{4.26} = 2.33 \text{ constrain to 1.0 per NFPA 68 (2013)}$$

$$E_2 = \frac{10^4 \times 1.63}{(1 + 1.54 \times 0.103^{4/3})202 \times 4.26^{3/4}} = 25.34 \text{ constrain to 1.0 per NFPA 68 (2013)}$$

$$A_{vf} = 1.63(1 + 1.18 \times 1.0^{0.8} \times 1.0^{0.4}) \times \sqrt{0.664/1.5} = 1.28 \text{ m}^2$$

By constraining E_1 and E_2 per NFPA 68 (2013) Para. A.8.5
a smaller required vent duct size occurs; therefore, keep the vent size at
$1.636 \text{ m}^2 \times 1550 = 2536$ in.

Summary of the Examples

Square/Rectangular Dust Collectors

In Part 1, the example dust collector, Figure 2.2, was reinforced to ensure that the vessel could resist a $P_{Red} = 5.0$ psi. In Part 2, Figures 14.1 and 14.2, the process was changed to corn starch and the resultant P_{Red} for the corn starch dust was found to be 3.55 psi for the square/rectangular Figure 14.1, and 4.29 psi for the square/rectangular Figure 14.2, both well within the original design pressure P_{Red}. The original explosion vent size 24″ × 24″ is subject to corrections to the vent and duct as follows.

Corrections

A_{v1}: L/D ratio was calculated, and it was found that there was no correction to the vent size required as the L/D was less than 2.0.

A_{v2}: The average air velocities were assumed to be less than 20 m/s; therefore, no correction was necessary.

A_{v3}: The mass of the explosion vent panel exceeded the threshold mass; therefore, the 24″ × 24″ original vent panel needed to be increased to a 36″ × 70″ panel.

A_{v4}: A partial volume correction was not applied.

Duct back pressure correction

The result of the back pressure analysis indicated that the 36″ × 70″ duct, 20 ft long with a 90° bend would not cause a significant change in the explosion flowing pressure P_{Red}. No additional correction was required.

Cylindrical Dust Collectors

In Part 1, the example cylindrical dust collector, Figure 3.2, was reinforced to ensure that the vessel could resist a $P_{Red} = 5.0$ psi. In Part 2, Figures 14.4 and 14.5, the process was changed to corn starch and the resultant P_{Red} for the corn starch dust was found to be 8.87 psi for the cylindrical dust collector Figure 14.4 and 5.51 psi for the cylindrical dust collector Figure 14.5. In both cases, the L/D ratio exceeded 2.0, and vent area corrections were necessary. New explosion vent sizes were calculated for each figure to reduce the P_{Red} values to 5.0 psi. For Figure 14.4, the required vent size is 28″ × 28″. For Figure 14.5, the required vent size is 24.6″ × 24.6″. The same corrections as calculated for the square/rectangular dust collector vents must be calculated for the cylindrical dust collector vents.

Note: The vent size represents a particular vent area and other vent size combinations including round vents are acceptable as long as the area is maintained.

For additional information, guidance, and examples, refer to NFPA 68 (2013).

Interconnections between separate pieces of equipment require special considerations and are addressed in NFPA 68 (2013) details.

16

Other Methods of Explosion Pressure Reduction

There are other acceptable methods of explosion pressure control than explosion relief vent panels.

Chemical Suppression System

High-rate discharge flame barrier uses pressure sensors that detect an explosion. When an explosion is sensed, this system discharges a cloud of sodium bicarbonate into the equipment. This discharge cools and extinguishes the explosion. It must be noted that this discharge induces a pressure in the vessel and must be considered when the allowable pressure of the system is calculated. Consult the manufacturer for details of operation. A disadvantage of this system is that the product would be contaminated, and a major cleanup is required. Initial cost versus maintenance and inspections must be considered when comparing a suppression system with an explosion vent panel installation.

Ten bar Rated Mills

There are some systems that are rated for 10 bar (145 psi) where the system can be isolated with 10 bar rated fast acting valves. If an explosion is detected by pressure sensors, the fast acting valves close and isolate the system to contain the pressure. Consult the manufacturers for details of cost and maintenance details.

Explosion Vented Equipment System Protection Guide, First Edition. Robert C. Comer.

Flameless Venting System

Flames are instantly quenched inside a box by efficient cooling. The burst pressure on the explosion panel of the box is 1.45 psi; however, the manufacturer must be consulted to determine if there is a resultant rise in flowing back pressure (P_{Red}) above the burst pressure of the panel.

In all cases of explosion relief, containment, or suppression, the effect of the resultant pressure in the system must be evaluated carefully, and a stress analysis of the equipment using Part 1 is performed to ensure that the system is safe.

Appendix B Part 2: Worksheet

Dust Collection System Checklist

System Identification____

Process Product____

Product P_{max}__________ Product K_{st}__________ P_{red}__________

Basis of P_{max} and K_{st}: Was product dust tested ______Y ______N

Were values approximated from a chart ______Y ______N

Other: ____

Dust collector: If square or rectangular

Pressure rating from manufacturer__________

Basis for rating ______% of 0.2% yield strength, or other__________

Is vessel located inside or outside?__________

Top width__________ Depth __________ Height __________

Top sheet thickness __________

Sides height __________

Sidewall thickness __________

Reinforcing on top or walls______________________________

Volume of dust collector________________________________

Dust collector: If cylindrical

Pressure rating from manufacturer______________________________

Basis for rating ________% of 0.2% yield strength, or other________

Diameter__________

Type of top: Flat______ Ellipsoidal__________ Torispherical__________

Thickness of top________

Thickness of cylinder wall ________

Length of cylinder ______________
Volume of dust collector ______________________________________
Hopper height ___________ Discharge size ______________
Hopper wall thickness __________
Dust collector/hopper flange size and thickness ___________________
Bolt size and spacing ____
Volume of hopper ____
Number of bags/cartridges ____________
Bag/cartridge diameter __________ Number of bags/cartridges ______
Bag/cartridge volume ______________
Access door width and height _____________________
Access door thickness ____________________________
Reinforcing on door ____
Type of latching:
Clamps________________ Bolts _________________________
Hinges ____________ Hinge pin diameter __________________
Nozzles: Diameter __________________________________
Dusty air inlet dimensions _____________________________
Clean air outlet dimensions ____________________________

Leg supports:

Size ___________ Number of supports____ Length of supports ____
Footpad size and thickness ______________________________-
Attachment to floor or ground_____________________________

Explosion vent information

Manufacturer/part number_____________________________
Size ________ Round _______ Square/rectangular ______________
Vent pressure rating____________ Vent area ____________________
Flange size _______________ Bolt size and number _______________
Distance from top of dust collector ______________________________

Duct information

Size________________ Length______ Wall thickness______________
Any bends at end of duct? ____________________________________

Screw conveyor

Cover width__________ Cover length___________ Thickness_________
Spacing and size of bolts or clamps on cover ______________________

References

ASTM E1226-19, Standard Test Method for Explosibility of Dust Clouds, (2019) ASTM International.

Baumeister (1958). *Mark's Mechanical Engineer's Handbook*, 6e, 11–72. McGraw-Hill.

American Institute of Steel Construction (1986). *AISC Manual of Steel Construction*, 8e.

Metals Handbook, 8e, (1961) , Volume 1. American Society of Metals

Chuse, R. (1977). *Pressure Vessels, The ASME Code Simplified*, 5e. McGraw-Hill.

NFPA 68 (2013). *Standard on Explosion Protection by Deflagration Venting*, 2013 Edition. NFPA.

Vallance and Daughtie (1951). *Design of Machine Members*. McGraw-Hill.

Young, W.C. (1989). *Roark's Formulas for Stress and Strain*, 6e. McGraw-Hill.

Explosion Vented Equipment System Protection Guide, First Edition. Robert C. Comer.

Index

e

f

g

h

Printed and bound by CPI Group (UK) Ltd, Croydon, CR0 4YY
16/04/2025
14658343-0001